Beyond Einstein:
Life, the Hyperverse and Everything

Beyond Einstein:
Life, the Hyperverse and Everything

Gary Warren

Hypervortex Publishing, 2014

First Printing: 2014

ISBN 978-0-9912697-0-9

www.hypervortex.com

garywarren@hypervortex.com

Cover by Tom Policano, tom.policano@ifotoz.com.

Dedication

To my wife,

who was there at the aha moment that underlies this work and who has accompanied me on the journey that led to its fruition.

Contents

Contents

Preface

Einstein contributed to multiple areas of physics, most notably quantum theory and relativity. Despite much progress, in the end he was dissatisfied, as were at least some of his contemporaries. Of quantum theory, Einstein's expressions of dissatisfaction have been paraphrased as "God does not play dice with the universe". Of relativity, Einstein's concern was even more serious. He proposed that the theory led to a paradox, which he called "Spooky Action at a Distance", a paradox that has since been confirmed. His last surviving significant contemporary, Dirac, who successfully predicted the existence of anti-particles, said of particle theories, that we do not understand even the electron.

I have long believed that revisiting Einstein's original approach and interpretation, and using data that has since become available, someone or some group, could achieve a new theory that is simpler, broader, more complete and more satisfying. I never really expected to complete such an effort myself, or even to make as much progress as is shown in this book. However, without the "publish or perish" pressures of academia, and with decades of exposure to the breadth of physics specialties that a PhD in physics and a long career as a Chief Scientist have allowed me, I have achieved a result worth sharing. The work resolves both of Einstein's concerns and also Dirac's concerns.

The new model of the universe presented here has been developed over the course of years. Portions have been presented and published. However most of it has been kept private until now, checking and rechecking that it overcomes various challenges that occur in so many other models of the universe.

All the checking and rechecking has confirmed, as well as I can, that the new model resolves the old challenges. It also reveals new physics with new challenges. It is tempting to spend more time continuing to check the correctness of the model. However, it will never be entirely correct. No matter what may have been achieved here, the future will bring refinements. That is the nature of man's effort to understand the universe. By publishing the model in this book, now a community of scientists can check and extend the model.

Thus, thanks to my friends who prodded me to birth this book from its private home on my computer out into the world. Here, without further ado is my new unified model of life, the hyperverse, and everything.

Introduction

The shared goal of physicists is to achieve:

A compact, physically meaningful representation of the essential properties of the universe from which all of its behaviors derive.

Measured against that goal, standard models of the universe, by their own admission, explain approximately 5% of the universe. In particular, they explain small scale things – elementary particles and things up to the size perhaps of our solar system. The standard models fail at the scale of galaxies and larger. Between the scale of solar systems and galaxies, the standard models explain many things but have substantial difficulties, such as with the detailed properties of black holes.

This work presents a model to explain the 95% of the universe that the standard models do not explain. Also, it provides a context for the standard models in the new model. It explains the existence of particles in the larger context, and provides place in the new model to insert the standard models. The new model unifies the essential properties of the universe into one equation that spans the smallest to the largest features.

The new model is a member of the class of models called fluid models. Such models have been important since medieval times. For centuries, various versions of a fluid called Aether were explored, with great success. Dr. Albert Einstein's early work attempted to understand the universe in terms of Aether. Development of fluid models continues in various forms.

The standard models attempt to explain the universe as being built from the bottom up. That is, they start with the smallest objects, and then develop the properties of larger objects as the sum of the properties of the smallest objects. When the models fail at the large scale, patches are invented that have no basis in the standard model. Standard models add an *ad hoc* "dark matter" to explain the difference between their model's results and the real world results for galaxies. Similarly, such models also add *ad hoc* "dark energy" to explain the difference between the model's predictions for expansion of the universe and the real world measurements of that expansion.

The new model, in contrast to the standard models, explores the universe top down. It starts with universe as a whole and drills downward to explore the formation of smaller objects as perturbations in the fluid, in particular as turbulence and vortices in the fluid. In the new model, the motions of an omnipresent fluid explain the largest scale behaviors of the universe. Also, in the fluid, are vortices. Vortices are common in fluids and are known in various fluids as eddy currents, maelstroms, tornadoes, hurricanes, and more. In the new model, galaxies are large scale vortices and the giant black holes that we observe at the centers of galaxies are the centers of those vortices. These galactic scale vortices spin off smaller vortices, which in turn spin off even smaller vortices. The properties and motions of the smallest of these vortices are what we observe as elementary particles. One may view the standard models as detailed mathematics of the observable properties of the smallest of these vortices.

The basic properties of the new model's omnipresent fluid are minimal and simple. Unique features of the new model, compared to the fluid model that Einstein and his predecessors hypothesized, include:

1. The hyperfluid's configuration, not the warping of space, causes our observation of a universe with complex shape.
2. The fluid has at least four spatial dimensions, in addition to a time dimension, and is therefore a hyperfluid.
3. Vortices in the fluid are what we observe as particles and the structure and behavior of the various vortices determines the properties of the various particles.

The above three features are essential to the model's success. Together, these features overcome the problems that befell Aether models. The first feature resolves the "Spooky Action" paradox that concerned Einstein and also simplifies the mathematics. The second feature is enabled by the first feature and is included because fluids with three spatial dimensions were studied fairly exhaustively long ago and could not reproduce Einstein's results for Special Relativity. The third feature is included because the alternative – particles composed of separate materials floating in the fluid – had been tried and failed. This third feature turns out to be so central to the model's ability to unify relativity, quantum theory, and particle theory, that it gives rise to the name of the model as a hypervortex model.

The need for a book, rather than a journal article, to present the model occurs in part because a hyperfluid, having more physical dimensions than we can observe requires substantial text to help the reader correctly visualize the physical model. The task is further challenged by potential confusion between the hypervortex model and other fluid models, including other hyperfluid models. Much text is expended throughout the book to help the reader internalize the correct hyperfluid representation, not some blend of it with other models.

The first seven chapters of the book are focused on helping the reader translate standard explanations of phenomena into the hypervortex model's explanations. They provide a top down introduction to the universe as a fluid. They introduce the fluid's properties and then drill down to show how it and the vortices that form within it create the various things that we observe.

The book presents the model mostly without the mathematics. The mathematical proofs are isolated to Chapter 8, which is written such that reading it is optional. Also, the complete mathematical foundation is compactly presented in a one page recap at the very end of the work. The mathematics chapter presents the one equation that unifies physics. From that one equation, a set of generally applicable equations of motion are derived, and from those arise the various specialized equations used today in various specialties of physics.

The reader may recognize many pieces of the hypervortex model from other sources – most of the pieces are from other sources. One or more scientist, or science fiction author, has postulated and published a speculation for just about every piece of physics. Physics has been described as a jigsaw puzzle to be assembled. The success of the hypervortex model is that its hyperfluid provides a unified picture into which the jigsaw pieces assemble. The effort to place each such piece in the overall jigsaw puzzle is a work in progress. Not every possible jigsaw piece ever hypothesized will fit into the new model. Yet it is surprising how many such hypotheses have already been successfully absorbed.

As the work continues, the need to refine the model is expected. Thus, the hypervortex model may generate variants. Indeed, the hypervortex model may expand to become a class of models. Both the equations and the approach used to develop the hypervortex model support such expansion. Various chapters discuss expansion possibilities in multiple contexts.

The value of a good unified model of the universe is more than its ability to unify known pieces. Rather, its value is largely about the model's ability to reveal previously unknown bits of knowledge. It would be egotistical of mankind to believe that mankind has already observed all of the universe's behaviors – all of the jigsaw pieces. The hypervortex model confirms that many pieces have not been observed. Chapters 9 and 10 present new phenomena revealed for the first time by the hypervortex model. These later chapters include possible tests of the model, possible refinements to the model, and the possibility of achieving previously impossible technologies.

1. Of Particles and Hyperfluid

The Michelson-Morley experiment is already considered a turning point in the development of modern physics theory. Adoption of the hypervortex model will make their experiment even more pivotal to the development of our understanding of the universe.

Prior to the Michelson-Morley experiment, the standard model of the universe included both fluid and particles. That fluid, which was called Aether, filled most of space. Particles filled the remainder of space. The particles and fluid were considered to be comprised of separate materials with distinct properties and roles in shaping our experience. The fluid supported at least the flow of light waves, radio waves, electric fields and magnetic fields. Particles formed the physical objects that we observe and which obey Newton's Laws. At least some people thought that the motions of the fluid generated gravity, though they did not explain how the masses of the particles participated in that action.

It occurred to researchers of that time that studying the boundary between the particles and fluid was both possible and valuable. Michelson and Morley therefore together built a speedometer to measure the velocity of the fluid at that boundary. The Earth itself, being a large collection of particles in motion as it orbits the sun, provided, it seemed, a perfect test case. As the Earth orbited the sun they measured the speed of the Earth through the fluid.

The speedometer of Michelson and Morley always measured a speed of zero. No matter where or when they placed the speedometer, their speedometer measured the speed of the Earth relative to the Aether as zero. Their measurement result created a conundrum. In their model, such a result meant that the fluid had viscosity. Viscosity is a fluid property that would cause the fluid speed near any particle to adjust to be zero relative to the particle. However, viscosity would also cause the particle's motion to be slowed, which would violate Newton's Law that objects in motion tend to stay in motion. Thus, their experimental results created a contradiction.

Einstein realized that he could advance our understanding of the universe via clever use of the results of the Michelson-Morley experiment. Specifically, Einstein leveraged the "relative" nature of the measurement result to develop his theory. His theory did not explain the experimental results, but rather learned from them. In particular, he noted that he could avoid the issue of Aether viscosity, instead dealing with the properties of light as measured by the Michelson-Morley experiment. Over time he promoted several relevant observations into postulates, including:

1. The speed of light is independent of the velocity of the emitter.
2. The principle of relativity; that motion of a reference frame cannot be distinguished by local measurement
3. Wave-particle duality

Using these postulates Einstein made some amazing predictions, since confirmed. Regarding relativity, he predicted that a particle's mass increases as its speed approaches the speed of light. He also predicted that clock rates slow down as their speed approaches the speed of light. While such predictions were amazing and correct, Einstein's Nobel Prize was awarded for other amazing work. That award recognized his correct predictions and explanations regarding light's behavior in the context of quantum mechanics, which he also derived from the above set of postulates.

What Einstein did not do is explain what was happening at the boundary between particles and fluid. Over time, lacking such explanation in Einstein's efforts, and with promotion by others of a geometric space-time interpretation of Einstein's results, Einstein's work on relativity became interpreted as describing properties of space itself. Such approach not only left the conundrum unresolved, it amplified one issue into a full blown paradox, the "Spooky Action" paradox described by Einstein himself. Theories in which relativity became tied to curved space-time made the speed of light an upper limit, in contradiction to a confirmed quantum phenomenon, predicted by Einstein, in which information is communicated between certain particles, entangled particles, much faster than the speed of light.

Since such mechanism contradicts the curved space-time interpretation of relativity, Einstein called it "Spooky Action".

In the century and more since the Michelson-Morley experiment, modern standard models have evolved that include particles, but mostly without fluid. Over the past hundred years, and more, these models have come to explain approximately five percent of our universe. To be clear, our understanding of particles has been dramatically improved over the past hundred years. However, explaining five percent of the universe is not the goal. Also, the standard models still have not been able to resolve the "Spooky Action" paradox exposed by Einstein more than half a century ago. And now, absent a fluid in the model, the particle models are adding dark matter and dark energy to the model, which may be reasonably viewed as surreptitious restoration of a non-particle component to the model.

Given the status of the standard model after more than hundred years of abandoning a fluid, this work returns to the turning point of physics to address the question not yet answered, "What happens at the boundary between particles and whatever else there is at the boundary?"

This work provides an answer. To date it is the only available reasonable non-magical answer. The answer that this work presents is that particles are not separate material from the fluid. Rather they are vortices comprised of an omnipresent fluid and existing in that fluid. This simple statement resolves the experimental results obtained by Michelson and Morley. As soon as one declares that the particles and the fluid are comprised of the same material, the boundary problem disappears. There is no longer a boundary between two materials. The Earth is comprised of fluid as is the space above the Earth. The speed of the fluid changes slowly and continuously at and near the Earth's surface. Michelson and Morley's speedometer is also made of fluid. Wherever the speedometer is placed the speed of the fluid in and near it is the same as the speed of the fluid that comprises it. Thus, the speedometer, which measures the difference in speed between the speedometer and the fluid, always measures zero.

The idea of particles as vortices in a fluid solves the boundary problem independent of the detailed properties of the fluid. The idea that fluid velocity changes without discontinuity within a fluid is generally applicable to every known fluid, even fluids with no viscosity. It is almost automatic. That is, it is hard to construct a reasonable fluid for which this is not true. This is because the velocity of fluids does not generally support discontinuities, i.e. abrupt changes in the velocity.

Since almost any fluid will address the boundary issue, the idea that particles are fluidic vortices opens up a whole area of research to determine the properties of a fluid that best unifies physics. The work presented in this book is largely such an investigation. It identifies and presents the results of such research in its current status. It looks for fluid properties that:

1. Satisfy Einstein's postulates,
2. Reproduce the known observed properties of electromagnetism and gravity, and
3. Provide observable properties of the universe that are otherwise attributed to dark matter and dark energy.

The idea that particles are vortices in the fluid provides additional benefits. First, it resolves the "Spooky Action" paradox. The paradox has existed because without fluid as a root cause of our observation of curved space, such curvature became assigned as a property of space itself. Such assignment made the speed of light the upper limit for the speed of anything. With fluid restored, curvature becomes a result of the properties of the fluid. As such, since fluids can support waves of various types at varying speeds "Spooky Action" simply becomes an indicator that, in addition to supporting the propagation of light waves, the fluid supports waves that travel faster than light.

A second additional benefit of the new model is that, at large scales, vortices in the fluid replace the need for dark matter. Dark matter is a concept invented to explain the large vortex-like structure of galaxies. A fluid with vortices explains such structure without a need for dark matter (or perhaps in another view, the fluid is dark matter).

Similarly, the fluid replaces the need for dark energy. The properties of galaxies and of the universe's expansion can be used to determine some properties of the fluid.

A third additional benefit of particles being vortices in the fluid is that the concept provides a real physical meaning to Einstein's concept of wave-particle duality. His version of the concept has always been vague. It simply stated that both particles and waves sometimes behave like particles and sometimes like waves. It gave no root cause and no specifics of when they behaved in each way. The hypervortex model adds a specific root cause for this duality – it says that both particle and waves have similar properties because they are both composed of the same material. Specific behavioral similarities are discussed in the upcoming chapters.

A fourth additional benefit of particles being vortices in the fluid is that the concept provides a more satisfying underlying physical explanation for the observed spectrum of particles. Fluids support a range of vortex types. Each corresponds to a particle type. This correspondence can be used to combine the standard particle models with the hypervortex model to create a complete model. For example, "strings" in modern particle models become short, observable segments of vortices in the hypervortex model.

A fifth additional benefit of the model is that by restoring fluid as a fundamental component of the universe we can dispense with complex curved geometry and we can restore the separation of the concepts of space and time. In the new model particle masses and clock rates change as their velocity increases because of properties of the fluid. As such, we can represent the fluid in a simple coordinate system and then determine the properties of the fluid that cause our observation of mass and clocks whose properties change as their velocity increases.

A sixth additional benefit of the model is that adding fluid to the universe reduces the number of types of matter required to explain the universe. Standard particle models claim and require the existence of many fundamentally different types of matter including leptons, quarks, gluons, photons, the W, Z, and Higgs bosons, dark matter, dark energy; and that list does not include composite matter types such as baryons, and mesons which are combinations of the more fundamental matter types. In contrast to particle models, the hypervortex model requires only one type of matter, hyperfluid. In the hypervortex model all other matter types are hyperfluid behaving in various ways.

The simplicity of the notion that particles are vortices in the fluid begs the question of whether the concept in some form was considered previously. The answer is yes, sort of, and this work benefits from those efforts. In particular, prior efforts showed that such solutions do not work if only three spatial dimensions are used. Thus, this model uses four spatial dimensions. The addition of a fourth spatial dimension was not considered acceptable during the era of the prior efforts. Now, particle models add multiple such dimensions. The addition of a spatial dimension opens up a whole new area of research of fluid models, which this work explores.

The spatial dimension added here is different than those added in particle models. In order for Einstein's work to succeed without an extra dimension, he needed to give the time coordinate properties of both space and time, which led to the concept of curved space which led to the "Spooky Action" paradox. To avoid that paradox, we split Einstein's time concept into two parts: one that is only a time coordinate and another that is only the spatial coordinate. This creates the additional spatial coordinate used in the model. The fluid here is called a "hyperfluid" to emphasize the presence of the additional spatial coordinate. Similarly, the vortices in this work are called "hypervortices".

The aforementioned benefits of the hypervortex model are more than sufficient to warrant restoration of fluid models as the mainstream approach to unification efforts. This book's direction is to explore those benefits, while also checking for potential problems with such models. The results of both aspects of the effort are presented in the following chapters.

2. The Basic Hyperverse

This chapter describes what appears to be the simplest fluid model that explains known science. Numerous variations have been considered, and if future experiments show preference for any of those variants, that is progress. Meanwhile, to date, the simplest identified fluid model that explains known science is a hyperfluid model.

The Hyperverse: The Hyperverse is a four dimensional (4D) ocean of hyperfluid. Actually it may have more than four spatial dimensions. However, in the hypervortex model there has been no evidence yet that our experience is affected by more than four. The laws of physics are the same throughout the Hyperverse. Since our best telescopes have observed neither an edge to the known universe, nor regions where the laws of physics vary; therefore the Hyperverse is a large ocean of hyperfluid with no apparent islands within the observable region of it. Whereas hyperspace is a mathematical construct, the Hyperverse is a physical place containing our universe and a near infinity of other universes. It is not a 3D ocean with a little bit of a fourth dimension. It is not a 3D ocean with space-time as a fourth dimension. It spans four spatial dimensions; time is separate.

The fourth spatial dimension is not different from the three that we observe. It only seems special because we cannot observe it. Also, it may seem special because for the past 100 years or so, much of the discussion about extra dimensions has been about dimensions that are special. However, in the hypervortex model the fourth dimension is not special. It is in principle observable. Currently, we usually observe only 3D because of certain details of the hyperfluid's motions, and the effects of those on us as observers. We are rather primitive observers on a cosmic scale. The hypervortex model tells us why we can observe only 3D, which enables us to begin to explore ways to observe the full 4D Hyperverse.

The Hyperverse is large enough to contain many 3D observable universes. The 3D universes in the Hyperverse are physically parallel to ours. Together, the 3D universes fill the Hyperverse. Every location in the Hyperverse is in one, and only one, 3D observable universe. The 3D universes are also temporally parallel; that is, events in nearby parallel universes are correlated in time. That is, they physically look much like our past and our future. The closer they are to our 3D universe the closer they are to being configured as our near past or near future. The 3D parallel universes are not physically separable from each other or from the Hyperverse. A single 3D observable universe cannot exist separate from the 4D ocean of hyperfluid. The reason why we observe the Hyperverse as if it is split into separate 3D observable universes is discussed later.

The location of the Hyperverse in a larger context is unknown. The nature of its borders is unknown. Whether it even has borders is unknown. There may be multiple hyperverses, multiple separate large bodies of hyperfluid. If so, any 3D universes in those hyperverses are likely not parallel to ours. While the basic properties of the hyperfluid may be the same in multiple hyperverses, the varying currents and environments in such multiple hyperverses will make the 3D universes different. The laws of physics require that neighboring 3D universes within any one hyperverse exhibit a degree of parallel evolution; they do not require parallel evolution of 3D universes that are each in different hyperverses. The properties of the Hyperverse in which our 3D universe resides are derivable, as it turns out, from our observations. In contrast, discussion of the properties of other hyperverses would be much more speculative at this time. For example, other hyperverses might contain hyperfluid that is different from the hyperfluid that fills our Hyperverse.

Basic Properties of the Hyperfluid: Hyperfluid fills the Hyperverse or at least the portion of it relevant to our experience. The space we call vacuum is full of this hyperfluid. Everything we are and everything we observe is determined by the properties of the hyperfluid. The most basic property of any fluid is that, unlike a solid, it can flow and rearrange itself. Three basic types of hyperfluid motion provide the foundation to explain our observations of the known universe. These basic types of motion are:

1. Laminar flow – smooth, unobservable motion of the whole, like a river moving uniformly and smoothly such that its motion is hard to detect.
2. Eddie currents, aka vortices and maelstroms – basically, structures in which the fluid flows around a center, like the maelstrom that Edgar Allen Poe describes in "A descent into the Maelstrom", or the small eddies that form when water drains rapidly from a tub.
3. Waves – oscillating hyperfluid motions, akin to transverse (ocean) waves that we see at the boundary between air and water, and also akin to the longitudinal waves that, in air and water, we interpret as sound.

The details of these motions are different in the hyperfluid than in water and air and they determine the nature of our observed universe.

The hyperfluid equations were derived in large part by using measurements taken across history to determine the details of the aforementioned motions of the hyperfluid. The derivation approach follows Einstein's initial direction for his work on relativity, except that he, and all others of that era, assumed a fluid (Aether) that fills three spatial dimensions. The work here extends that effort and adds a spatial dimension because doing so completes the task.

In an ocean of water, we can directly observe vortices, and waves, but not the laminar flow, at least not from inside the flow. This limitation regarding the laminar flow is fundamental and applicable to all fluids. It is also central to the concept of relativity – wherever we are in the fluid, motion is RELATIVE to the laminar flow. Objects that we see in the hyperfluid are manifestations of vortices within the hyperfluid. Such objects range in size and type from elementary particles to galaxies. The detailed properties of the vortices determine the detailed properties that we measure for the various objects. Among and between those vortices is the laminar flow that we cannot see, and which we therefore call space, deep space, or the vacuum of space. Transverse waves in the hyperfluid are interpreted as light. Evidence supports the existence of additional wave types that create rare but important behaviors.

In many local regions of space, the hyperfluid has a relatively uniform laminar flow. The direction and speed may vary in different regions. Einstein's amazing success with his theory of Special Relativity was, and is, essentially, that he derived the effects that the laminar flow have on our observations without the ability to observe such flow itself. His three spatial dimensions are the three spatial dimensions perpendicular to the laminar flow. Einstein's mixing of space and time approximates the effect of viewing events in one region from a point of observation in another region where the laminar flow is in a different direction. His equations also describe the effect of moving between regions whose laminar flow is in different directions.

To Visualize the Hyperverse: If the reader is at this point trying, but failing, to visualize three spatial dimensions perpendicular to a fourth, the reader likely grasps the information being presented. For much of the upcoming discussion it will be sufficient to visualize a dimension along the direction of the laminar flow and a select two of the three dimensions perpendicular to the direction of the laminar flow. The third of those perpendicular dimensions can be omitted or retained like a vestigial appendage dangling from the image. Importantly, the difficulty of visualizing the dimension is not at all at odds with its existence. Also a quick read of the short 19th century book _Flatland, A Romance of Many Dimensions_ may help, not with visualizing, but rather with understanding the difficulty of visualizing it. Alternatively, one might review the history of public interest in that book as the realization has grown over the past century that such a dimension might truly exist. In the past, man had difficulty visualizing the Earth as a sphere rather than as flat, and also visualizing it as other than at the center of the universe. We will find ways to visualize the Hyperverse, and of hyperfluid in motion within it.

The laminar flow of the hyperfluid is the reason we can only observe three dimensions. Briefly, the laminar flow causes vortices to align with the flow. That alignment causes all transverse waves to travel perpendicular to the laminar flow. That alignment also causes our sensors, and our senses, to align perpendicular to the laminar flow, which is convenient because it allows us to observe the transverse waves. However, it also means that we cannot observe anything along the direction of the laminar flow – our sensors are not looking in that direction, and there are almost no waves to observe coming from that direction. Ergo, we observe only the three dimensions perpendicular to the laminar flow.

To help visualize man's overall situation in the Hyperverse, imagine a colony of genius germs floating at night in a maelstrom on the surface of the ocean. Our germs have no ears and so cannot hear sound waves (longitudinal waves) propagating from air above or from the water below. The genius germs can sense the surface waves of the ocean (the transverse waves) via the accelerations that the waves cause the germ to experience. Our colony of genius germs carefully analyzing the waves might realize that the colony is in a vortex. Also, our germs might observe beyond that vortex the presence of other vortices, and they might even deduce a curvature to the surface between vortices. Further, our germs might write equations for all that is observed, and such equations would all assume two dimensions. Their two dimensional equations would reference a curved space whose curves follow the surface of their local piece of ocean. Finally, if one of those germs realized and presented a description their true status as living on a 2D surface moving in a 3D universe, all of the germs, including the author of the new truth, would have great difficulty visualizing the third dimension.

Man today is like our hypothetical genius germs, except man writes equations for a three dimensional surface. Also, that surface is actually inside a volume comprised of one four-dimensional hyperfluid, not a two dimensional surface on the boundary between two different three-dimensional fluids – air and water. Like our germs, man is unable to observe the longitudinal waves, at least in any scientific manner to date. Thus, like the genius germs, man relies on the transverse waves for his observations. The vortex in which man resides is the Milky Way, and the other large vortices that man observes are other galaxies. Like the germs, man deduces a curved surface between galaxies. Because we call the transverse waves in the hyperfluid "light waves" therefore we declare that our observable universe bends to follow the path traveled by light. And, yes we will all have great difficulty visualizing the fourth spatial dimension.

Detailed Properties of the Hyperfluid: To support known observations, the hyperfluid is at least approximately an ideal fluid, a simple, pure, super-dense, non-dissipative, continuous, compressible, isotropic, classical hyperfluid. "Simple" here means that it has no internal structure or sub-elements; more about that later. "Pure" means that the hyperfluid does not contain other materials within it. At each point in space there is, in sufficient approximation to explain our known observations, either the hyperfluid or nothing. "Super-dense" means like the densest black hole. "Non-dissipative" essentially means frictionless; the fluid has no ability to convert motion into heat, or other such energy sink. "Continuous" means that quantities of hyperfluid cannot simply disappear and reappear. It also thus means that in order for fluid to get from one location to another the fluid must move along some path between a start and end point. "Compressible" means that the hyperfluid's density can vary; it can be increased by applying pressure to it. The hyperfluid's density varies across the Hyperverse. "Isotropic" means that the properties of the hyperfluid itself are the same in all directions. "Classical" means that the hyperfluid obeys Newtonian laws of motion rather than Special Relativity. For example, its speed is not fundamentally limited by the speed of light. The hyperfluid's motions are complex and governed by rules applicable to all fluids that have the properties assigned here to the hyperfluid.

Hyperfluid is, apparently, not composed of atoms or elementary particles, or other such substructure. Such a claim may seem strange in modern science, but before atoms were discovered, people were fine with water being composed of water and air being composed of air; and many people then would likely have thought anyone an idiot who asked what water was composed of. Now, it may be that the hyperfluid is composed of smaller particles, but to date some data suggests otherwise and no data seems to require it. For example, phenomena are consistent with hyperfluid that has no property of temperature, i.e. no thermal energy. Since temperature is typically inherent to the existence of substructure, the lack of a property of temperature implies a lack of substructure. In principle, the fluid may have some amount of friction or dissipative ability; however, if so, the amplitude of such effects is so small as to be ignorable for known physical phenomena. Also, if it does have dissipative properties then it might indeed also have a temperature, and then the possibility of internal structure would have to be revisited.

The hyperfluid does not have properties of mass, spin, or electric charge. Such properties, which we observe in particles, all derive from the behaviors of the hyperfluid. In particular, particle mass, spin and charge all form as a result of the behaviors of vortices that form in the hyperfluid.

There may be some contaminants in the hyperfluid. Any flotsam, as may be in a fluid, if present, appears to be ignorable for known physical phenomena. That is, the model describes known phenomena without regard to the possible presence of impurities. The hyperfluid may or may not be reactive with other materials; to our limited knowledge it is not exposed to other materials that would create observable reactions. It may or may not undergo phase changes that make it a solid or a gas. No known observations as yet require that the hyperfluid undergo phase changes, however it is possible that some circumstances might create a phase change. Exploration of the possibility of a gas phase is discussed in a later chapter. Circumstances in which the hyperfluid might violate continuity are discussed in that same later chapter.

Hyperfluid Flow and the Expansion Current: Hyperfluid is to the Hyperverse as water is to an ocean. Just as oceans have large scale currents, so too does the Hyperverse. The behavior of the Hyperverse is dependent on these currents. The dominant hyperfluid flow in our Hyperverse is the Expansion Current. The Expansion Current is a flow of hyperfluid along the extra spatial dimension, the fourth spatial dimension. Its effect is not observed directly, yet it is that flow that causes observers to note that the universe is expanding.

The Expansion Current makes the extra spatial dimension special. Because the hyperfluid itself, as a material, is isotropic, hence if there were no fluid flow, the observed properties of the universe would be the same in all of the spatial dimensions. Those observations might also be uninspiring for scientific exploration. Without the Expansion Current to drive the creation of complex motions in the hyperfluid there would be much less to observe in the Hyperverse. On its own, the hyperfluid tends to exist at a uniform density and without motion. The Expansion Current disturbs that static state. It creates a dynamic, changing environment whose complexity creates our existence and also creates the layers of physical phenomena that we struggle to understand.

The direction of the Expansion Current varies gently across the Hyperverse. Also, local currents can combine with the Expansion Current to vary the direction of the overall local current. The three spatial dimensions that can be locally observed by man are the three dimensions perpendicular to the direction of the overall current. Thus, across the Hyperverse, the varying direction of the hyperfluid flow forms the three spatial dimensions that we observe into a curved subset of the full four dimensional space; a 3D observable expanding universe. The parallel 3D universes are each a slice through the Hyperverse perpendicular to that large scale current. Our 3D observable portion of the universe is a curved moving, continuous 3D subspace in the Hyperverse.

Vortices in the Hyperfluid: The Expansion Current creates vortices in the hyperfluid. A vortex is a slightly more technical term for an eddy. Flow in any fluid has a laminar (i.e., smooth) component and a turbulent component. It is the turbulent component in any fluid that creates vortices. The hyperfluid flow is mostly laminar, but with some turbulence. Tornadoes and hurricanes are examples of vortices in air; the jet streams that circle the Earth near the North and South poles are examples of long lived vortices. The eddy current that forms at a drain when water flows out of a tub is a vortex. Vortices in super-fluid, as in Helium 3 super-fluid, are a less well known example, but with special value in the future for examining vorticity in the hyperfluid. This is because a superfluid is a 3D fluid that has no viscosity, i.e. no internal friction which may therefore help us to better understand vortices in hyperfluid which also has no viscosity. In all of these examples, the vortex is a structure comprised of the fluid in which it resides. A vortex is not a separate thing in the fluid; it is a dynamic structure with some degree of stability which gives it the prolonged existence that allows its observation and study.

Vortices in the hyperfluid, in some ways, are perhaps most like the jet stream or the eddy current at the tub drain. They have long length in the direction of motion of the fluid and rotate around an axis that exists along that length. Analogously, since the Expansion Current flows along the extra spatial dimension, the vortices have their length also along the extra dimension. We do not observe that length; we observe a very small slice through the vortex, essentially a cross-section of it, which is why we observe a 3D universe rather than a 4D universe – more about that later.

As with vortices in other fluids, vortices in the hyperfluid tend to have reduced density at their center, on their axis. Their density in fact appears to go to zero on the axis for a large class of vortices in the hyperfluid. The reduced density at the axis creates a force that tends to move fluid toward the axis. That force is counter-balanced by centripetal forces created by the rotation of the fluid around its axis. That balance creates the stable vortex structure.

There is a limit to the analogy between vortices in the hyperfluid and vortices in air, water, and other typical fluids. Specifically, vortices in the hyperfluid can be billions of light-years long. The ratio between the vortex length and its radius is much higher in the known vortices of the hyperfluid than in the vortices we otherwise typically observe. Also, where the radius of a tornado, for example increases as the air in the upper atmosphere thins, the same effect can happen to vortices of the hyperfluid. However, in the hyperfluid the effect is stretched out over billions of light years. The change in the radius of a hyperfluid vortex over the course of a human lifetime is negligible. We do not see the length of these vortices; the length occurs along the dimension that we cannot observe. When we observe a vortex we observe only the radius of a cut across the vortex, which is much smaller than the length of the vortex.

The detailed density and velocity profiles of various vortices are determined by certain detailed properties of the hyperfluid. In particular, vortex profiles are determined by the hyperfluid's response to variations in its density and its velocity. Vortex profiles determine the strength of the various forces that we observe among particles as well as the observed properties of individual particles – properties such as mass, electric charge, and spin.

Particles in the Hyperfluid: Each particle that we observe is the intersection of a vortex and our observable 3D universe. The real corresponding object in the Hyperverse is the vortex. We cannot observe the whole vortex, only the piece of it that intersects our 3D universe. Vortices generally cross through many observable universes; each vortex is observed as a particle in each universe through which it crosses. This creates the parallel structure among neighboring3D universes.

The detailed relationship between the vortex structure and the velocity of the corresponding observed particle creates our observation of Einstein's Special Relativity. The length of vortex intersecting any 3D universe varies as the observed velocity of the observed particle changes. At higher observed velocity, the length grows. Also, the angle of the vortex relative to the overall Expansion Current changes as the observed velocity of the observed particle changes. When the particle is at rest, the vortex axis is parallel to the Expansion Current. As velocity increases, its angle relative to the Expansion Current increases; at the speed of light its angle is perpendicular to the expansion current.

Most vortices generally cross through any particular observable universe once. Some cross through twice. In such cases, they create two coupled particles. We cannot see the coupling because the remainder of the vortex, including the part of the vortex that is between the two particles, resides outside our observable universe.

This coupling between two such particles is the basis of Einstein's prediction of "Spooky Action" also known as entanglement.

Forces on Particles Due to Vortex Interactions: Every force that has been ascribed to particles actually happens to the corresponding vortices. Vortices in fluids interact and change each other's motion, or even each other's structure. This can be seen among vortices in a turbulent body of water. It can be seen among hurricanes at times when multiple exist simultaneously in a region. Also, the jet stream interacts with hurricanes to affect their motion. The same occurs in the hyperfluid. We observe four forces: gravity, electro-magnetism, strong and weak forces among particles. Each force is a result of a particular aspect of the properties of vortices.

Gravity: Gravity is our observation of the fact that, in the hyperfluid, objects are attracted to regions where the fluid density is reduced. The vortex of any particle that has mass creates a density well, i.e., a gravity well. Things tend to "fall" into that well. The force law for gravity is determined by the shape of the gravity well. In most models, the properties of gravity, such as its relation to mass and distance are added as assumptions into the model. In contrast, in the hypervortex model the properties of gravity are derivable from the properties of the hyperfluid.

The size of a gravity well scales in proportion to the total energy of the corresponding vortex. We observe gravity as proportional to mass because, by Einstein's equation $E=mc^2$, mass is proportional to the vortex energy. The size of the vortex is proportional to its energy, which is proportional to the reduction in hyperfluid density which is proportional to the induced gravity. Thus, historically we think of mass as causing gravity, but the hyperfluid shows that it does not. Gravity and mass are correlated observable phenomena both caused by the presence of a vortex whose hyperfluid density is appropriately reduced compared to the hyperfluid density in the surrounding free space.

Electromagnetism: Electromagnetic forces are our observation of the effects of rotational motion of the hyperfluid near a vortex. The direction of the rotation determines whether the electric charge is positive or negative. At any location in the vortex, the magnitude and direction of the electric and magnetic force is determined by the local variations in the rotational speed of the hyperfluid. When two vortices are parallel, we observe a strictly electric force between the vortices. When two vortices are not parallel that creates the observation of magnetic forces in addition to the electric forces.

The relation between vortex orientation and the forces it generates is a consequence of how our observation of relativity occurs in the hyperfluid. In the hyperfluid, two vortices with the same orientation relative to the Expansion Current travel at the same velocity, as do the corresponding observable particles. From standard electromagnetic physics, for two particles traveling at the same velocity we can calculate the force between them using a reference frame where their speeds are both zero. In such a reference frame there is no magnetic field between the particles. Conversely, when their orientations are not parallel, there is no reference frame in which both vortices have no speed. The motion generates observable magnetic forces. Multiple prior hyper-dimensional models have recognized the same fact. The advantage of the hypervortex model is that it unifies electromagnetism with all the other forces. It also clarifies the appropriate choice of reference frames for various calculations, as discussed in a later chapter.

Strong Forces: For some particles the shape of its vortex follows a simple form such that the corresponding gravitational and electromagnetic forces obey the expected mathematical form at all distances from the axis of the vortex. Whether very near the axis or very far from it, the forces created by such particles follow the expected behavior for gravity and electromagnetic forces. Such particles have no strong force. Other vortices have complex shapes near the axis that cause anomalous interactions with other vortices that approach into that physical space. We call most of those anomalous interactions the strong force. Its details are very complex not unlike of the forces that would occur between two hurricanes in close proximity.

Weak Forces: When vortices become so close to one another that their presence starts to affect the structure of the vortex axes, their proximity creates the possibility that the vortices will destabilize and change form. We observe this as the creation and destruction of particles, and we call this particular anomalous interaction the weak force. Particle models provide a mechanism for the weak force but no root cause for its existence. The hypervortex model shows that the weak force exists because the hyperfluid density of vortices goes to zero at the axis, at least for particles that have mass. Thus, two such vortices cannot be in the same place at the same time. That is, no second massive vortex or structure can be at the very axis of such a vortex because there is no fluid to manifest that structure. Thus, in such situations, the interaction between the particles can cause the axes to restructure, which can change the very nature of the vortices.

Waves in the Hyperfluid: Fluids support various different wave motions. Air and water both support sound waves at high speed relative to the speed of other waves in those fluids. Also, at the interface of air and water we can physically see a second type of wave – the waves that are formed by the wind blowing and as the wakes of boats. This type of wave travels much more slowly than sound waves. Vortices in air and water support a third type of waves – waves that travel along the length of the vortex. Transverse motion of the jet stream – as in its oscillations as it moves north and south bringing warm and cold weather to regions – is wave motion along the length of a vortex. Similarly, waves along the length of tornadoes from the ground to the upper atmosphere are waves along the length of vortices.

Hyperfluid supports each of the three wave types that we observe in air and water. However, the hyperfluid's properties are different from those of air and water and those differences make its waves a bit different, both qualitatively and in terms of their speed. The hyperfluid is super-dense yet soft and compressible. Since wave speeds are higher in dense materials; hence the speed of waves in the hyperfluid tends to be much higher than the speed of the corresponding wave type in water or air.

Light is a type of wave in the hyperfluid. It is the only type of wave in the hyperfluid that we easily observe. We do not directly observe the other types of waves in the hyperfluid because we have no sensors for them. Light corresponds, not to sound waves in air or water, but rather to the much slower moving surface waves on water. Still, the hyperfluid is so dense that these slow waves in hyperfluid travel much faster than sound waves in water. The fact that light is a slow wave in the hyperfluid is important and key to the hyperfluid's success explaining certain phenomena that particle models cannot – more about that later in the chapter.

Light waves have a type of structure in the hyperfluid that lacks a density well. Rather a photon, like a very small wave on water, creates a very small temporary oscillating perturbation in the hyperfluid's properties at any location as it moves by. Because, of this, multiple photons can be in the same place at the same time. Just as small waves on water can pass through each other, photons in the hyperfluid can pass through each other as they move in different directions. Also, photons can join together as they move in the same direction; Lasers are an example of individual photons of light joining together to form into a larger wave.

Like a wave on water, a light wave's oscillation is in a direction other than its motion. A wave on water oscillates up and down while it moves horizontally. Also, just as waves on water follow a 2D surface in a 3D body of fluid; light waves follow a 3D observable universe in the 4D Hyperverse. It is this property of light waves that divides the hyperfluid into separate 3D universes – more about that later.

Light waves do not generate gravitational fields. Gravitational fields are caused by density wells, i.e., central voids at the axis of particle vortices where there is no hyperfluid; however photons have no density wells – no mass. Thus, just as waves on water do not create a force that attracts boats; light waves in the hyperfluid do not attract particles. Despite this, light waves are still subject to the effects of the density wells present in other vortices. Just as waves on water bend around islands, piers, and other obstacles in the water, light waves in hyperfluid bend around regions, such as the density wells of vortices, where there is a lack of hyperfluid. We observe this bending in part as diffraction (which is exactly the bending of light around objects). We also observe it as the gravitational effect of particles on light waves. The gravitational effect is the reason that light waves bend and change direction when passing through the gravitational fields of stars. Their change in direction is the same as that experienced by particles in a gravitational field. But whereas particles are also accelerated as they fall into a gravity field, light waves do not go faster when headed directly into a star. Just as waves on water cannot be sped up or slowed down by pushing on them, light waves cannot be sped up or slowed down by the forces that affect particles or their vortices. Still, just as the speed of waves on water is affected by the depth of the water, the speed of light waves can be affected by the local properties of the hyperfluid.

Consciousness and Observing 3D Universes: The hypervortex model provides a very different view of the existence and nature of 3D universes than particle models. In the particle models each 3D universe is spatially separate. A typical description, in the particle models, is made by analogy to a basket of hyper-apples. In that basket, the skin of each apple is a 3D universe; the space between the apples is not discussed. Also, the space within each apple is not discussed.

The hypervortex model replaces such *ad hoc* construction with the Hyperverse full of hyperfluid. As described earlier, each 3D universe is a subspace of the Hyperverse that is locally perpendicular to the Expansion Current and to the vortices in it. That means that each subspace bends, as discussed later. Now, the reason that we observe only one such subspace has two main parts. First, light waves are emitted perpendicular to the vortices. Second our consciousness is fundamentally light-like, which is explained below. These two factors together explain most of why we observe only three dimensions of a larger space.

Light waves can be emitted from any point along the length of a vortex. However, once light is emitted, it follows a path that is perpendicular to the expansion current. Thus, each 3D subspace of the Hyperverse is, approximately, a subset of space containing all of the light waves that can interact. These subspaces have some thickness along the extra spatial dimension (along the direction of the expansion current) since photons have some thickness in that dimension. Also, if we try to imagine each 3D universe as truly separate, we will find that many photons exist that overlap two neighboring 3D universes. The subspaces are not truly separate, but the physical similarity of nearby subspaces is so high that we observe these differences only at the quantum level, as discussed later.

Because our consciousness is light-like, our consciousness, like light, stays in the subspace in which it was created. And thus our consciousness only observes light emitted in that subspace. Our bodies are composed of vortices that exist simultaneously in very many of these subspaces; our bodies exist in more than billions of these subspaces.

Each subspace is one observable universe and each consciousness in each subspace exists within a 3D cross-section of the vortices that comprise our bodies. If our consciousness were fundamentally corporeal we would observe four dimensions or more, and we would observe a full 4D instance of our bodies. Whether we would recognize our 4D body is an interesting question. Another interesting question is whether we can cross between observable universes. Actually the question applies to our consciousness; our bodies are already also in other observable universes. It appears that there may be ways and times where consciousness may cross into another observable universe basically by sliding slightly along our 4D body, enough to shift out of one observable universe and into another. This raises another interesting question. If this happens, would we notice? The nearby subspaces will appear very similar to our own, and be equally populated with consciousness in each living body. This and related questions are discussed in more detail in a later chapter.

If our consciousness were fundamentally corporeal we would not observe relativity. We observe relativity because our consciousness is light-like. Light is fundamental to what we observe because our consciousness is fundamentally light. This logical conclusion follows from using Einstein's work as he wrote it – distinguishing observers and their measurement tools (clocks and rulers) from reality. Einstein's derivations distinguish the nature of observation from the nature of reality. Unfortunately, the particle models, the geometric models, took Einstein's results as being reality. That is a fundamental divergence between the hypervortex model and mainstream particle models. Particle models do not include a concept of an observer into the model as a factor that affects the results of observations. When the distinction between observation and reality is restored into the model development process, then fluid models succeed. Further, then the hypervortex model provides knowledge about the nature of consciousness. Particle models cannot provide knowledge of the nature of our consciousness because those models do not have the concept of us, the observers, as a component in their models.

Observing Curvature, Expansion, and Relativity: It was stated above that light curves to remain perpendicular to the Expansion Current. This occurs because of the interaction of light waves with the expansion current. Further, based on observation, as well as solutions to the hyperfluid equations, each 3D universe curves to close on itself as the 3D surface of a hypersphere. Thus, the parallel 3D universes form into one nested set of 3D spheres, one inside the other. (The Hyperverse could perhaps be ovoid rather than specifically spherical.) This nested set of 3D spheres is Billions of light years across. Our expanding universe is one of these spheres amidst a near infinite set.

Particle models also include 3D universes as closed surfaces on a hypersphere, which is why those models describe a universe as the surface of a hyper-apple. However, in particle models, each 3D universe is the surface of a separate hypersphere, which is why they describe parallel universes as the skins of a basket of hyper-apples. The idea of nested spherical universes is not entirely new for the hypervortex model. Such a construct was around already in the 1970's and then described as a hyper-onion. Each layer of the onion was analogous to an observable 3D universe. Further, such a model was based on an "ideal cosmological fluid", and was quite popular at the time, before the now standard quark models took hold. However, because those efforts did not keep Einstein's distinction between reality and the observer, the developers of those models could not get the detailed mathematics to work. The standard models now make each universe a surface of a separate sphere because the mathematics of nested spheres still does not work in their approach. It does not work because they still do not distinguish between reality and the observer.

In the hypervortex model, the distinction between the observer and the reality is restored and thus the mathematics of relativity works for nested 3D spherical shells forming a hypersphere. As the Expansion Current flows, the vortices move such that an observer in any particular 3D universe observes every particle as obeying Einstein's special relativity; the vortices move such that every observer in very 3D universe observes a particle that obeys it.

No "Big Bang" Singularity and No Dark Energy: The spherical shape of the Hyperverse corresponds to an Expansion Current that emanates from the center of a hypersphere. Further, it means that each 3D spherical surface in the hypersphere is expanding and all of them are expanding in lock step. The density of the hyperfluid declines as the Expansion Current flows outward. The speed of light and the speed of clocks become reduced. Also, the size of the cross-sections of vortices increases. The details determine the observed size of particles, the expansion velocity and any observed acceleration of that expansion. And, the measured changes generally differ from the actual changes.

The standard models have something mathematically similar to the Expansion Current, but without the corresponding physical concept. They also, again, do not have the distinction between reality and the observer, and thus omit the expansion's impact on the rulers and clocks. Thus, those models compute a special time during an assumed "Big Bang" when physics was different. Also, they hence need dark energy as a patch to explain the observed expansion acceleration. That is, the predictions of the particle models do not match the known data, however, rather than discard the particle models, they invent a concept they call dark energy whose sole purpose is to fix the discrepancy.

The hypervortex model changes our understanding of the evolution of the universe over time. First, the distinction between reality and observer enables realization and that at the supposed moment of a big bang at some zero time, observers there would have observed a universe much more similar to what is observed today than is predicted by standard models. In particular, the rulers and clocks scale so that the observer at that time would not see any singularity. In some versions of the hypervortex model, the clocks and rulers scale such that any observer at any time anywhere in the expanding universe will measure the same age and size of the sphere that we measure.

The hypervortex model provides simple mechanisms to explain a changing expansion rate of the universe. For example, the expansion rate will vary if a gradient in the hyperfluid density is causing a real accelerating Expansion Current or if the changing measurement scales cause an apparent acceleration. Either way, the hyperfluid needs no dark energy to explain an accelerating expansion.

No Need for Dark Matter: Particle models assume that all of the particles in a galaxy collectively generate enough gravity to hold the galaxy together. However, measurements show that there are insufficient particles in our galaxy to provide the required amount of gravity. Thus, particles models hypothesize dark matter as an invisible matter of an unknown type that provides the additional mass required to create the gravity. That is, particle models are wrong about the root cause of the structure of galaxies and dark matter is their patch.

In contrast to the particle models, the hypervortex model needs no patch to explain galactic cohesion. Our galaxy looks like a vortex and in the hypervortex model it is a giant vortex. The galactic vortex is so large that the entire galaxy exists within it. In the hypervortex model the galactic vortex provides the gravity that holds all the stars in the galaxy. In the hypervortex model, the fact that there is insufficient particle matter to hold the galaxy together is proof that each galaxy exists as a super-large vortex in the hyperfluid, and that the particles within them are trapped there by the galaxy itself. Regions of turbulence in the Expansion Current have evolved over time to form clusters of vortices. Thus, clusters of galaxies have formed where the expanding universe had initial turbulence.

Past, Present, and Future as Correlated Universes: The superiority of the hypervortex model becomes apparent when examining the structural parallelism among observable 3D universes. The hypervortex model and some variants of the standard particle models include parallel temporal evolution of 3D universes. However, in such variations of the standard models, with 3D universes as skins of hyper-apples, there is no mechanism provided to exchange information among the apple skins to enable them to evolve in any correlated manner. This is a critical deficiency in such models. Those models are built with the speed of light as a fundamental limit, but they need much higher communication speeds to create the parallelism among spatially separated 3D universes. In contrast, in the hypervortex model, each vortex penetrates whole sets of adjacent 3D universes as the mechanism enforcing correlations. In the hypervortex model no special mechanism is required to create correlations among parallel universes.

In the hypervortex model, the Expansion Current interacts with vortex motion to cause 3D universes to be correlated in a manner as past, present and future. If one 3D universe is specified as the "present", then all the 3D universes on one side are similar to its past, in varying degrees, while all the 3D universes on the other sides are similar to the future, in varying degrees. The smaller the distances between a 3D universe and the one defined as the "present" the more similar their configuration. The correlation, however, is not 100%. They are not actually the past and present. Each observer stays in one 3D universe as its configuration evolves. The correlation occurs because vortices have long length and thus penetrate many 3D universes. De-correlations occur due to various detailed effects on vortex motion. These effects include, for example, quantum effects, relativistic effects, weak interactions, and free will. The details of such effects are discussed much later after all of these topics have been introduced and discussed individually.

The Properties of Elementary Particles: The hypervortex model and some of the particle models share a common basic description of the most basic physical feature of elementary particles; specifically, in these models particles have length along an extra spatial dimension. In string models, particles have been found to be strings, having some small length along some spatial dimension. In the hypervortex model, the particle is the portion of the vortex that is within the observable 3D universe. That portion is very similar to, or perhaps identical to, the string in string models.

In the hypervortex model, the particle mass is proportional to the vortex's energy per unit length multiplied by the length of vortex in a 3D universe (and the proportion is given by Einstein's famous equation, $E=mc^2$). Likewise, in the hypervortex model, charge, spin, and other measured properties of the elementary particles all result from aspects of the vortex properties, and depend on the vortex type and size.

In the hypervortex model, the lifetime of a particle is related to the length of the corresponding vortex. The vortex travels through each observable universe at the speed of light. When the end of a vortex passes through a universe the particle is observed as spontaneously decaying in that universe. Thus, a vortex that is a billion light-years long will be observed as a particle with a lifetime of a billion years, or so. (Artificially created particles have additional factors that may affect the observed lifetime.)

As was mentioned earlier, the sign of the electric charge of a particle corresponds to the direction of rotation of the vortex. Whichever direction of rotation corresponds to a positively charged particle, a negatively charged particle rotates in the opposite direction. The property of particle spin is not directly related to the rotation of the hyperfluid around the vortex axis. Rather it relates to a secondary rotational effect. Notably, the property of particle spin, in the hypervortex model, is due to vortex features that are similar to historical expectations from which the property name arose. The term "spin" was developed at a time when physicists' models were much imbued with the concept of fluid as a root cause for physical phenomena.

In contrast to the hypervortex model, the standard models mostly lack physical root causes for the various particle properties. Unlike the hypervortex model, the standard models have no root cause for particles being strings. Even the historical root cause for the name "spin" does not apply in the standard models. The standard models do, however, have robust mathematics to describe the relations among the many common and rare elementary particles.

An effort to combine the root causes from the hypervortex model with the particle mathematics of the standard models has barely begun, yet the prospects for synergy seem good. A key remaining question regards quarks; is each quark a vortex or does each quark represent a feature of a vortex?

Anti-Particles: Anti-particles are our observation of the intersection of "anti-vortices" and our observable universe. For each type of vortex, the corresponding "anti-vortex" is the same except its symmetry is reversed. That is, the "anti-vortex" is like the vortex seen through a mirror. For example, the anti-vortex rotates around its axis in the opposite direction compared to the vortex. In the natural world anti-vortices are not usually observed, because the vortices are created by turbulence in a region of space and all (or the vast majority) of vortices acquire the same symmetry as the symmetry of the overall turbulence in that region of space. This is a known property for vortex formation in fluids in general. For example, we observe this on Earth with tornadoes and hurricanes. All the tornadoes and hurricanes in the Earth's northern hemisphere rotate in the same direction. This is not due to any fundamental nature of the universe. Rather, the vortices acquire their direction of rotation from the direction of the rotation of the Earth. In the southern hemisphere the tornadoes and hurricanes rotate in the other direction.

When we create a vortex artificially in the hyperfluid, however, as in particle collisions, we can and do create some particles with opposite symmetry – anti-particles. The artificial creation mechanisms allow their symmetry to be determined by factors unrelated to the symmetry of the local region of space. For example, when we create pairs of particles from energy (i.e. light), we create a particle and its anti-particle – a vortex and its corresponding anti-vortex. This occurs because there is no initial turbulence and, by creating the vortex and anti-vortex together, the total net turbulence is still zero. Anywhere in space, the natural creation mechanism selectively creates particles with predominantly the same symmetry, but we can create particles and anti-particles.

Basic Quantum Mechanics: The hypervortex model shows that quantum mechanics relates to two phenomena that occur along the length of vortices: (1) waves that travel along the length of vortices and (2) a tendency for vortices to reconstruct their long length if broken into fragments by an obstacle.

The presence of waves traveling along the vortices reduces our ability to predict detailed vortex motion. That, in combination with our ability to observe only a very short piece of any vortex that is in our universe at any time limits our ability to know the properties of waves that travel along the vortices. The resulting knowledge deficiency creates uncertainty when computing the motion of particles and outcomes of events. Quantum mechanics is in part a way to compute the possible range of particle motions and the probability distribution of those motions despite the knowledge deficiency.

The tendency of vortex fragments to reassemble creates several related phenomena. For example, consider a vortex passing through a region of space which would block the path of some particles and provide multiple different paths for various other particles to traverse. Entering such a region, a long vortex will split into shorter fragments. The various fragments take different paths through that complex region of space and, upon exit, reassemble. As a result the vortex has sampled the whole region of space and behaves differently than is predicted by non-quantum mechanical theories. Other examples and more details are provided in the chapter specific to Quantum Mechanics.

"Spooky Action": "Spooky Action" is a known effect in which an event in one place affects an event elsewhere in a way that requires information to travel much faster than the speed of light. First predicted by Einstein, "Spooky Action" exposes a fundamentally non-statistical aspect of the properties of the waves that propagate along vortices and thus a fundamentally non-statistical aspect of quantum mechanics. Einstein proposed an experiment as follows: Create two particles via some event such that their combined value for some physical property – specifically a property for which quantum mechanics says the value for each particle individually is probabilistic (e.g. particle spin) – must have a particular outcome. After their creation, the particles travel some distance apart before measuring that physical property of those particles. According to quantum mechanics, each particle acquires a particular value for that physical property when the measurement is made. Thus, to achieve the required combined value when one particle is measured, the other particle must instantly take on a particular value. To do this, Einstein noted, some "Spooky Action", some instant communication, must occur. Such experiments have been performed and confirm the existence of this "Spooky Action".

Einstein proposed the experiment as a way to expose a contradiction that exists in most models, including his own. Specifically, "Spooky Action", as faster-than-light communication, contradicts models that assume or predict that nothing can travel faster than light. He had discovered the fatal flaw of the models that had been birthed from his work and was trying to get the work redirected into a direction that would remove the contradiction. Yet, even today, the contradiction is inherently ignored in all models that assume that nothing can travel faster than light. That includes all the standard models. Consequently, the standard models have been completely unable to unify quantum mechanics and gravity.

The hypervortex model resolves the contradiction. First, the model indicates that the particles created in Einstein's proposed experiments are two observations of a single vortex (or sometimes other structure in the hyperfluid). Whatever event creates the two particles actually creates one vortex that extends over time from the creation point outward in both directions. However, the Expansion Current pushes most of that vortex into adjacent universes over time. That vortex passes through our 3D universe twice. In between, it is in other nearby 3D universes.

We observe a particle at each of the two places where the vortex passes through our 3D universe. Rather, or more specifically, we see a particle and its anti-particle because the vortex passes through our observable universe in one place with the same symmetry as all the other particles of that type in that region of space and it passes through the observable universe with the anti-symmetry in the other place. However, that detail is an aside in this discussion. Rather, the important point about the two particles as regards "Spooky Action" is their being two observations of one vortex – because it is the vortex that provides the link that supports "Spooky Action".

A wave propagating along the vortex provides the necessary communication to achieve the "Spooky Action" between the two particles. When the first particle is measured and enters a particular state, the measurement generates a wave that propagates along the vortex. The wave traveling along the vortex has a much higher speed than light waves traveling within our observable universe. The wave traveling along the vortex quickly reaches the other particle on the same vortex and puts it into the corresponding appropriate state. This does not violate Einstein's Special Relativity. Special relativity's rule that limits the observed rates of motion applies to light waves that travel within the observable universe and to the observed motion of the intersection of the vortices and the observable universe.

3. Basic Motion in the Hyperfluid

This chapter qualitatively addresses the hypervortex model's ability to explain and generate basic conservation principles including Newton's Laws. Its statements are backed by the mathematical rigor found in Chapter 8. This chapter is important, though perhaps a bit pedantic. Still, it qualitatively shows how the hypervortex model provides deeper understanding than is provided in particle models.

Basic Conceptual Differences: The hypervortex model has three basic conceptual differences relative to the current mainstream models. These differences, together, distinguish it both from particle models and from older fluid models that predate Einstein's work. These differences, collectively, create a new class of physics model that has a chance of succeeding where other models have failed to achieve a fully unified model.

The first conceptual difference is that in the hypervortex model particles are not fundamental. Fluid is fundamental. Particles are not separate objects that flow in the fluid. Particles are structures comprised of fluid. This concept is not entirely new. While all of the standard models and most of the prior fluid models treated particles as being a fundamental aspect of nature, Lord Kelvin (of the Kelvin temperature scale) conceived of the possibility that particles were like smoke rings in a fluid. The hyperfluid, to a degree (no pun intended), picks up on his concept and develops it.

The second conceptual difference is that the hypervortex model restores Einstein's distinction of observation from reality. Einstein's special relativity showed that our clocks and rulers, which are the basis of our observations, are affected by motion. In his original interpretation, to understand reality, we had to apply a correction to the measurements to determine what was really happening. Later, when relativity became interpreted to mean that space and time changed, that change removed the separation of observation and reality. In the now standard interpretation of special relativity any change in the clock is automatically interpreted as a change in the actual rate of time. In such interpretation there is no universal time.

As a consequence of distinguishing observation and reality, the hypervortex model restores the separation of measurements from reality. In the hyperfluid, clocks and rulers are made of particles, which are made of fluid. Motion changes behaviors of the fluid which changes the behaviors of clocks and rulers, but not time and space. To understand what is actually happening in time and space we must correct for the effects of the hyperfluid's behavior on the outputs from clocks and rulers. We know that such corrections can be made because such corrections are made regularly in practice to synchronize high precision clocks (such as those inside the satellites that give us GPS) whose rates are affected in ways that Einstein's theory described. Such corrections are used to map the high precision clocks to a universal time. Correspondingly, the hypervortex model includes universal time and local clocks whose rates vary.

The third conceptual difference between the hypervortex model and most other models and is the separation of space-time into space and time as separate concepts. Before Einstein's work space and time were separate concepts. For most of us, these concepts still are separate. The mathematics of his special relativity includes aspects that have been interpreted as a coupling of space and time. Here, the hypervortex model uses four spatial dimensions and time. This allows determining whether time and space are fundamentally coupled, or whether Einstein's approach led to a coupling of space and time coordinates due to an insufficiency of spatial dimensions in his assumptions. After all, he assumed that there were only as many dimensions as we observe. If there are more dimensions than Einstein assumed, as many variants of today's standard models now claim, then Einstein's approach should be reapplied with additional spatial dimensions. The hypervortex model does this. In other models, four dimensions means three spatial dimensions and time. In this model four dimensions means four spatial dimensions; time is separate.

Inertial Motion of Particles: Newton's law of inertia says that things in motion tend to stay in motion and that things at rest tend to stay at rest. More precisely, it says that in the absence of forces on objects, the objects' velocity stays constant. During Einstein's lifetime, physicists determined that particles, not comprised of fluid, but moving in a fluid could not both satisfy Newton's law of inertia and the observed constancy of the speed of light. The mainstream fluid models of that time were such models, and thus failed. To overcome this failure, most models since that time have assumed the existence of fundamental particles and not a fluid. However, all of that history ignored the concept of particles as structures comprised of the fluid. Particles comprised of fluid can behave in accord with Newton's Law of Inertia.

Observation of Newton's Law of Inertias is used to determine likely properties of the hyperfluid. Newton's law of inertia is satisfied if the hyperfluid is an ideal fluid that has regions that are free of turbulence. Also, satisfaction of Newton's law of inertia requires that the fluid have properties that support the existence of stable particles in motion. Stable vortices in motion in an ideal fluid tend to stay in that motion anywhere in which the fluid's properties are uniform. This is true of almost all, or perhaps all, possible ideal fluids that satisfy other very general properties of fluids. Since particles are the observed portions of vortices, inertial motion of vortices leads to our observation of particles obeying Newton's law of inertia. It has not been formally proven that the hyperfluid must be ideal in order to support behaviors that match Newton's Laws. Thus, it is likely that the law of inertia requires an ideal hyperfluid, but some non-ideal hyperfluid scenarios might also support our observation of inertia. The hypervortex model, as presented here assumes an ideal hyperfluid, while allowing variant models with variant sets of properties.

An ideal hyperfluid enables particles to obey Newton's Law of Inertia because such fluid cannot internalize the momentum or energy of any vortex or wave. It lacks an ability to absorb energy as heat. It lacks any ability to dissipate energy into properties such as temperature, or entropy. That is, it has no friction. Without friction, hyperfluid creates no net drag or friction, on a vortex to slow it down.

Quasi-Inertial motion of vortices can be observed in Earth's weather. Tornadoes and hurricanes are vortices akin to electrons and protons, respectively. When other winds are not applying forces to tornadoes or hurricanes, they move in relatively straight paths and at relatively constant speed. Air is compressible fluid like the hyperfluid except it is not an ideal fluid. Thus, the behavior of the tornadoes and hurricanes is only quasi-inertial. The air can absorb energy from them and hence their motion and their existence can be dissipated if new energy is not supplied.

Electrons and protons obey laws similar to the tornadoes and hurricanes except that electrons and protons are vortices in a four-dimensional fluid. We observe only a three dimensional particle because we only observe the part of the vortex where it intersects our observable universe. (Why this is true is discussed later.) In the analogy to tornadoes and hurricanes, our ability to see only a 3D particle from the vortex is akin to observing the tornado or hurricane where it touches the ground. The intersection of the tornado or hurricane and the ground forms a circular 2D spot. The intersection of a vortex in the hyperfluid with our observable universe forms a spherical 3D spot, that we call a particle.

Non-Inertial Motion of Particles: Non-inertial motion is when objects change their velocity. This may change their speed, their direction of motion, or both. In almost all or perhaps all fluids, non-inertial motion occurs when and where a vortex enters a region where the fluid properties are non-uniform. This applies in the hyperfluid not because of any special properties of the hyperfluid but rather because it is a rather typical ideal fluid. We do not however, directly observe the non-uniformity of the hyperfluid properties. Rather we observe objects changing velocity and ascribe the cause of these accelerations to forces on the objects. By definition, forces are phenomena that cause objects to change their velocity. The root causes for our observation of the various forces are the local behaviors of the hyperfluid.

Hyperfluid behaviors that can create forces on a vortex include waves nearby that vortex and other nearby vortices. In general, any non-uniformity of any aspect of a fluid can create a force on a vortex upon its entering the region of non-uniformity. The force on a vortex will continue for the time period that it remains in the region of non-uniformity. In the hypervortex model, so far, there appear to be only two properties that can be non-uniform. These are hyperfluid density and the hyperfluid motion. All observed forces are caused by non-uniformity of these two properties of the hyperfluid. There is nothing atypical about variations in fluid density and fluid motion causing forces on vortices. Tornado and hurricane motion is changed by variation in air density and air flow. The detailed relation between the observed forces and the variation in the hyperfluid density and motion are perhaps unique to the hyperfluid. Chapter 8 explores mathematical representation of a hyperfluid whose properties give rise to the observed forces. Other representations that produce the same, or sufficiently similar, predictions to match measured data may be possible.

Particle Collisions: A special case of non-inertial motion occurs when objects scatter off one another as in a physical collision. Physical collisions are events such as two cars colliding. Most of us take for granted that objects cannot pass through each other, but physics models must explain this. Particle models of the universe say that, ultimately, the reason particles collide, as in a head on collision, rather, than pass through one another is that the particles are Fermions. While that explanation is true, it is more or less tautological. Fermions are, by definition, particles, of which there cannot be two in the same place at the same time. Therefore by definition, in the particle models, such particles cannot simply pass through each other. Instead they affect one another's behavior in a way that we call a collision. Particle models provide substantial phenomenology correlating the Fermionic property to other properties of particles; however they do not provide a physical root cause for this Fermionic property of the particles.

The hypervortex model provides that missing root cause for Fermionic behavior. In particular, the hypervortex model says that the observed collision outcome is the direct result of vortices getting so close to one another that they are directly affected by the region of zero fluid density at the core of each vortex. Where there is no fluid, there is nothing to support the structure of the vortex. Just as tornadoes and hurricanes can only exist in air, not in the vacuum of space, similarly vortices in hyperfluid can only go where there is hyperfluid. Since there is no hyperfluid at the center of each particle, hence the other particle cannot exist there. This is the root cause of their Fermionic behavior. Since the vortices cannot simply pass through each other, they behave as in a collision; sometimes they combine to form a new structure, even a new vortex.

Analogies to Fermionic behavior of vortices in air and water have been hard to find. Hurricanes and tornadoes tend to repel one another, but generally that effect is due to interaction of their rotational motions. The mutual repulsion of hurricanes with one another and of tornadoes with one another is more analogous to two electrons that repel because they have the same electric charge. Fermionic behavior is exposed when vortices approach each other with enough energy to collide despite deflection forces created by their rotary motion and where the density of the fluid at the axis of the vortex goes to zero. Perhaps such effects can be made to occur with vortices in Helium 3, a real ideal fluid in which the fluid density at the axis of vortices goes to zero.

Particle Creation and Destruction: When particles collide with enough energy, the particles may be structurally altered or destroyed. In the process new particles may be created. Experiments confirm this. However, where we see particles being altered, created or destroyed, the reality is that vortices are being altered, created or destroyed. Just as a hurricane can spin off a tornado, vortices can similarly split off other vortices. In the process the total rotational motion of the vortices is conserved. That is, the total amount of rotational motion entering the collision is the same as the total exiting the collision. Similarly, the total amount of vortex energy is the same entering the collision as it is exiting the collision.

The more energy the particles have when they collide, the more energy and rotational motion the corresponding vortices have when they collide. The more energy and rotational motion there is in the collision, the more change may occur altering, creating, and destroying vortices. Other properties of the vortices, other than energy and rotational motion are similarly conserved during such collisions. When particles are caused to collide as in particle accelerators that are used to search for the Higgs boson, the conservation of various vortex properties is observed as rules about particle creation and destruction. The study of particle interactions has largely been an effort to identify and categorize these rules. The hypervortex model shows that the underlying cause of these rules is the fact that the particles are vortices in a fluid. With the development of the hyperfluid, the effort to identify and categorize rules can now be used to determine certain detailed properties of the hyperfluid.

Behavior that approximates the behavior of particle vortices can be seen in the creation of tornadoes and hurricanes. It is not the same as what we see in particle colliders, in part because in particle collision experiments we are creating an artificial circumstance that tends not to happen in the atmosphere. In any case, where weather fronts collide or where wind turbulence otherwise exists, the turbulence may form into tornadoes. Similarly hurricanes form where air flow has cross currents. In both cases at the collision or location where the wind patterns collide, there is rotational motion. It is that rotational motion that forms into the tornadoes and hurricanes, while conserving energy and rotational motion. Tornadoes and hurricanes are two types of vortices in air and each exists in various sizes. Similarly particle vortices exist in a limited number of types and each come in a number of sizes. Experiments that explore the collision of particles confirm this. Because the hyperfluid is four dimensional while the Earth's atmosphere is barely three dimensional, there are more types of vortices in the hyperfluid than we observe in Earth atmosphere. There are also more properties of the vortices that are conserved during collisions that create and destroy particles.

Wave Particle Duality: Einstein observed and stated the concept of Wave-Particle Duality. He observed that light and particles both sometimes behave like a wave and sometimes like a particle. He postulated therefore that particles and waves have properties of both, which is the essential point of the concept of wave-particle duality.

The hypervortex model explains wave particle duality in terms of the similarities and differences among various physical structures comprised of hyperfluid. The description of wave particle duality provided here is much more specific than that provided by mainstream models.

In the hypervortex model, the fundamental physical difference between photons of light and vortices of elementary particles is that the hyperfluid density goes to zero at the axis of particle vortices, while photons form structures in the hyperfluid that have no such void at the axis. That one difference gives rise to the observed differences summarized by the concept of wave-particle duality.

The particle-like behavior of both particles and light, according to the hypervortex model, occurs because particles and light both comprise localized structures of hyperfluid. Both light and particles are comprised of hyperfluid. Both are structures that have stable solutions to the same non-linear equations that couple the density and the motion of the hyperfluid. Stated in terms of observed forces, that coupling is a coupling between gravity and electromagnetics. For both light and particles, because of the non-linear coupling (most readers can ignore the word "non-linear" without substantial loss of meaning), both light and particle type structures in the hyperfluid have a minimum standard size specific to each type of structure in the hyperfluid. That size and specifically the compactness comprise a particle-like property. The minimum unit of light is a "photon". The word "light", incidentally, here refers to the full spectrum of electromagnetic waves, including visible light, all the radio waves, microwaves, x-rays, etc. For all of these types of light, the minimum unit of light is a photon, a particle-like compact wave. To get lasers and useful radio signals many photons are combined. Similarly for particle type structures, the smallest – electrons and protons (and maybe neutrinos an interesting detail for the future) – each is a compact structure in the hyperfluid.

The light-like behavior of both photons and particles occurs, in the hypervortex model, because both light and particles have structures that are long in one dimension and much smaller in the other three dimensions. Along the length the structures have periodic structure that we observe as waves. Also, being made of hyperfluid, the structures are not rigid; they are flexible. That flexibility enables them to support the existence of vibrations that we observe as waves, again, mostly along their length. For photons, we see wave structure easily because the photon is long in a dimension that is one of the three dimensions that we observe. However, for particles, the long length is in the dimension that we cannot observe, so we rarely see that wave behavior. The reason that photons orient their length differently than particles is in part because they have no vortex in their structure, and in part because they are emitted perpendicular to the particles.

Fermions and Bosons: Particles and photons behave differently in other ways, besides the fact that one mostly behaves like a particle, while the other mostly behaves like a wave. In particular, while, as discussed earlier in the chapter, typical particles behave like "Fermions", photons do not. Photons rather are observed to behave like "Bosons". "Bosons" are, by definition in particle models, particles that tend to be more co-located than statistically expected. The hypervortex model provides a specific explanation for the Fermionic and Bosonic behaviors in terms of their physical structure. Moreover, the explanation provides new insight.

First, photons do not have Fermionic behavior because they have no vortex and hence no region that is void of hyperfluid within them. Photons have regions where the hyperfluid density is slightly higher than the ambient hyperfluid density and other regions where the hyperfluid density is slightly lower than the hyperfluid density. The differences are very small however, perhaps a deviation of one part in a billion relative to the ambient density. Thus, when two photons are on collision course there is nothing to stop them from passing right through each other.

The lack of a Fermionic behavior in photons does not automatically mean that photons have Bosonic behavior. Photons are Bosons because the minimum sized structures in the hyperfluid tend to combine lengthwise to make longer structures. Structures in the hyperfluid tend towards configurations that have less "action". The term action has specific meaning that is discussed in the mathematics chapter. Action correlates with turbulence in the laminar flow. Less action correlates with less turbulence. Both particle and photon structures tend towards configurations that are stable moving structures. They are more stable with less action when they have longer length. Also, photons are more stable when they add in phase with each other, thus tending to create multi-photon structures with long length and more amplitude than single photons. We observe all of this as boson behavior.

The new insight provided by the hypervortex model regarding Fermions and Bosons is that particles are Bosons as well as Fermions. Specifically, they are Bosons along their length and Fermions in the other dimensions. Particles like electrons and protons nominally have very long length – billions of light years. However, many scenarios cause them to break into pieces along their length. The important point here is that once the cause of the breakage is past, the pieces reassemble into their nominal long lengths. They do this for the same reasons that photons do. These breakages and re-assemblages are not directly observable to us, but we do observe the effects as quantum phenomena, the details of which are discussed later.

Mass: Another difference between particles and photons regards their mass. There are two aspects to mass – gravitational mass and inertial mass. Einstein assumed that the gravitational and inertial masses of a particle are equal, and hence we assign only one mass to a particle. The hypervortex model shows why these two masses are equal, although, since that discussion requires mathematics, it will not be provided in this section. Rather, only the source of gravitational mass is discussed here.

Particles have gravitational mass while photons do not. Particle models attribute this to the Higgs boson, which those models say is associated with an invisible field that is part of the infrastructure of the universe. The hyperfluid may be that invisible field. In any case, in the hypervortex model it is the vortex that again creates the distinction between the particles and photons. In the hypervortex model, the vortex at the center of particles causes the average density of the hyperfluid to be lower in a particle than in nearby free space. This causes the particle to have a gravitational field around it – gravity is a force created by gradients in the hyperfluid density. In contrast to the particle structure, the average hyperfluid density within a photon is unchanged relative to ambient hyperfluid. The hyperfluid density varies within the photon; however, without a vortex, its average density is the same as that of ambient hyperfluid. Therefore it generates no observable gravitational field and thus no mass.

Photon Creation and Motion: Because the presence of light does not change the average local density of the fluid, creating light is easier than creating particles – there is no excess hyperfluid to be removed. For the same reason, creating photons requires much less energy than creating particle vortices.

Photons are observed to be created by all, or at least almost all, simple non-inertial motions of a particle. In the hypervortex model, those photons are emissions from the particle's vortex. That is, as the motion of particle vortex changes, it emits light waves. Light waves so formed are very much like the wake of a boat (or perhaps the winds that stray from a hurricane). However, the analogy of light to the wake of a boat is limited in that a boat creates a wake even for inertial motion of the boat. Inertial motion of the boat creates a wake because water is a non-ideal fluid that is constantly decelerating the boat. That deceleration causes the formation of a wake that propagates away from the boat, and it is the reason that boats require a continuous supply of power to maintain the boat's speed. The water creates drag that decelerates the boat, emitting a wake; the engine applies a force that re-accelerates the boat. When the water's drag and the boat's engine output are equal we observe a boat moving at constant speed, but the wake is still present – the lower the drag on the boat, the smaller the wake.

To the extent that has been observed in experiments, particles emit light as a wake only when the particle vortices change their motion. This fact implies that the hyperfluid is an ideal fluid that creates no drag on the particles. There may be some small wake created even during inertial motion of particles – a wake too small to have been observed by current experiments. If such a wake is observed in the future, then reevaluation of whether hyperfluid is an ideal fluid would become appropriate.

Photons, being the wake of particle vortex motion, are emitted perpendicular to the vortex that creates them. The length of a photon's structure lies within an observable universe because it is emitted perpendicular to the particle vortex. Just as the speed of a boat's wake is independent of the speed of the boat, likewise the speed of light is independent of the speed of the particle vortex. Just as the wavelength of a boat's wake is dependent on the boat's motion, likewise the wavelength of light is dependent on the particle vortex's motion.

Structurally, light has a degree of similarity to waves on water. Waves oscillate the water vertically, while they move horizontally. Similarly in light waves the internal motion is perpendicular to the direction of travel of the light wave. Also, just as we see waves that are traveling in different directions pass through one another and continue on their path, photons can do the same. Waves on water are quasi-two-dimensional on the boundary of two three-dimensional fluids (air and water). Light waves are quasi-three-dimensional in the bulk of a four-dimensional hyperfluid.

For waves in water the average depth of water in a region is not changed by waves whose height is small compared to the depth of the water. Water waves can grow until their height is similar to the depth of the water. When they do, their behavior changes – they crash on the shore. Similarly, light waves in the hyperfluid can grow in amplitude until the localized changes that they make to the density of the hyperfluid is similar in amplitude to the total density of the hyperfluid. That is their amplitude limit, and when it is reached the behavior of the light changes. Such limiting behavior is seen in some very powerful lasers.

It may seem tempting to conceive of a model of the universe in which the 3D observable universe is the surface at the boundary of two 4D fluids (e.g. like the boundary between water and air). It may even be that such a model is more accurate than the one presented here.

However, such a model requires certain refinements to Galileo's definition of the principle of relativity and to modern extensions of it. Such changes are a challenge that has not been required to date to achieve a theory of everything. Perhaps the possibility of the observable universe as the boundary between two 4D hyperfluids can be explored in the future. Defining an experiment whose measurement will determine which view is more correct will be interesting, as will its execution. Meanwhile, Ockham's razor, a principle used in physics to compare and select models based on their simplicity, suggests the use of a hyperverse filled with one hyperfluid, not two hyperfluids. It is the simpler model and no need for a second hyperfluid has been identified. Hence it is the hypervortex model presented here.

Conservation of Hyperfluid: The overall quantity of hyperfluid appears to be conserved at least as regards physics local to our existence. It may be possible to convert hyperfluid to energy or to otherwise create or destroy it. However, so far no known observed physics effect implies such event. Explanation of the observed creation and destruction of particles, and of light, does not require any change in the overall quantity of hyperfluid. All such events are, as yet, explainable in the hyperfluid without any change to its overall quantity.

In principle, hyperfluid could be entering or leaving the Hyperverse through a boundary. Just as water enters oceans from rivers and departs via evaporation yet overall the ocean levels remains relatively constant, similarly hyperfluid may be entering and leaving the Hyperverse at various locations. Indeed, just as ocean levels have changed over long period as ice ages come and go, the density of the hyperfluid may change over long periods. The expansion current could be evidence of such change. Still, in such case the evidence to date is that any net change in the amount of hyperfluid in the Hyperverse results from a difference between the amount entering the Hyperverse and the amount departing.

In principle there may be a variety of hyperfluids with different properties and there may be different hyperverses composed of one or a combination of hyperfluids. The hyperfluid that comprises our Hyperverse may react with other hyperfluids if exposed to such hyperfluids. However, to date, no known observed physics has been identified as requiring multiple hyperfluids. That is, the hypervortex model succeeds (explains known phenomena) using a model of the Hyperverse as being filled by a single hyperfluid. If new observations violate the single fluid model, then the possibility of a multi-fluid should be revisited.

Conservation of Hyperverse Energy: Conservation of energy is an important aspect of the observed physical world. However while particle models of the universe use this concept, such models do not explain why energy is conserved. In contrast, the hypervortex model adds a root cause for conservation of energy, and it does so without need to introduce any new concepts. In particular, quite simply energy is conserved in the Hyperverse simply because energy is conserved in ideal fluids that meet certain basic properties. Multiple mathematical theorems regarding energy show that ideal fluids conserve energy in particular circumstances. The presence of the fourth spatial dimension does not change the theorems. Per the theorems, energy is conserved if the quantity of fluid is conserved and if there is no exchange between the fluid and its boundary. Since these criteria are met in the Hyperverse, or rather to the extent that these criteria are met in the Hyperverse, within our known observations, therefore the energy of the hyperfluid is conserved. The original source of the energy is unknown. Perhaps it all started as energy of the Expansion Current.

As in any physical system, the total quantity of energy in the hyperfluid is equal to the sum of the kinetic and potential energy. Also, as in any system, energy can and does slosh between these two forms. It is the total that is conserved. These two forms of energy exist in all fluids and in all systems. This is a standard aspect of all physical models. Certain detailed aspects of kinetic energy and potential energy in the hyperfluid are unique to the hyperfluid and they determine the laws of physics that we observe. The total energy of the Hyperverse is the sum (or integral) over all of space of the kinetic and potential energy at each location in space. Hyperfluid itself is not included as a form of energy; any energy content it might have is not relevant until such time as its quantity might become a variable. New observations of new details may provide insight for addition of new energy terms to the hyperfluid.

Kinetic Energy of the Hyperfluid: The kinetic energy of anything is defined as the energy contained in its motion. Thus, the kinetic energy of the hyperfluid is the energy contained in its motion. As in any ideal compressible fluid, the density of its kinetic energy at each point in space is a function of the fluid's density and its speed. The denser it is and the faster it is moving the more kinetic energy it has.

We observe properties of kinetic energy in air and water. Hurricanes are rated on their potential to do damage based on the speed of their winds. Wind is air in motion. The greater the wind speed the more kinetic energy it has and thus the more damage it can do. Tornadoes have even higher speed winds at their centers than hurricanes so they have more kinetic energy at their centers. Thus, the center of a tornado does more damage, for its size than a hurricane. Water is almost one-thousand times denser than air. Thus, water at any speed has much more kinetic energy than air at the same speed. Even at much lower speeds water has much more kinetic energy than air. A tsunami is much more destructive than a hurricane even though the water moves much slower than hurricane winds, because even the slow moving water has more kinetic energy than hurricane winds.

The hyperfluid is much denser than water. It also moves much faster. Most of that kinetic energy is the energy related to the expansion current as the hyperfluid moves in the direction perpendicular to our observable 3D universe. This current provides an enormous source of energy that provides the energy needed to generate galaxies and all the particles in them. A small amount of kinetic energy of the expansion current can be converted into energy sufficient to create galaxies.

The relation between hyperfluid speed and its kinetic energy follows classical, not relativistic rules for fluids; our observation of special relativity is not due to any fundamental property of the hyperfluid; it is due to the same mechanisms that limit our senses to the observation of a single 3D universe. The speed of the hyperfluid is not limited to the speed of light; it is shown in Chapter 8 that the speed of the hyperfluid determines the speed of light.

Potential Energy in the Hyperverse: The potential energy of the hyperfluid is energy of all types other than kinetic energy. It is energy that can be converted into kinetic energy, and back, without loss. At each point in space there are two types of potential energy in the hyperfluid. One form of potential energy in the hyperfluid is due to variations in the density of the fluid. This form of potential energy is akin to, but not the same as, the potential energy of air or other fluid under pressure. Where there is a variation in the hyperfluid density, the hyperfluid wants to flow from the denser region to the less dense region. This creates motion of the hyperfluid, and it changes the density of the hyperfluid. The conversion of energy between fluid motion and fluid density variations conserves energy. That is, there is no loss of energy in the conversion.

The second form of potential energy in the hyperfluid relates to variations in the fluid velocity. This form of potential energy seems to have no analogy in the properties of water or air. In known fluids, viscosity, a form of friction, rather than any type of potential energy, limits the magnitude of the velocity variations of the fluid. In most fluid if there is a local variation in the fluid velocity, that causes kinetic energy of the fluid to turn into heat of the fluid, warming it and slowing the motion of the fluid. However, hyperfluid has no temperature. In the hyperfluid, velocity variations provide potential energy that can increase the kinetic energy of the hyperfluid.

Perhaps it is the uniqueness of this property of the hyperfluid that has made the completion of a fluid model of everything a bit harder and with a longer history than might otherwise have occurred. The exact quantified relation between the size of the density and velocity variations and the quantity of potential energy is a key factor that determines certain details of the properties of the forces that we observe in nature. Chapter 8 quantifies that relation.

In principle, there may be other components of the potential energy; if so they create behaviors that we have yet to map to observable phenomena and so we cannot add them to the model.

Mass-Energy Equivalence: Einstein's equation, $E=mc^2$ generalized the principle of energy conservation by adding a notion that mass could be converted into energy, and vice versa. This is called mass-energy equivalence. Prior to his results, it was believed that mass and energy were conserved separately. His equation changed the meaning of energy conservation such that it is the combination of mass and energy that is conserved. In particle models of the universe it means that there are three types of energy, mass energy, kinetic energy and potential energy; and it is their sum that is conserved. Particle models however do not explain the equivalence of mass and energy. Einstein's equation quantifies the equivalence, but does not explain it.

In the hypervortex model, the concept of mass-energy equivalence derives directly from the fact that the vortices that we observe as particles are composed of hyperfluid. In the hyperfluid, mass energy is not a distinct third type of energy. Rather mass energy is the combination of kinetic and potential energy that is used to create the vortex. The kinetic energy of the vortex is the energy used to create the rotational motion of the hyperfluid in the vortex. The potential energy is energy used to create the hyperfluid void at the center of the vortex, and the associated variations in the properties of the hyperfluid between that central void and the space outside the vortex. Einstein's equation for mass-energy equivalence implies that the observed particle mass is proportional to the energy in the corresponding vortex. In particular, his equation says that the observed mass of a particle must be proportional to the energy contained in the portion of the vortex that resides in the observed 3D universe. Each vortex requires an amount of energy per unit length of the vortex. Further, each vortex type requires a different amount of energy per unit length. Thus we observe a different mass for each type. Such properties of vortices in hyperfluid are not necessarily true for all possible hyperfluids. Substantial effort has been required to determine detailed hyperfluid properties that satisfy this proportionality. The effort's success is a key aspect in which this book's content extends prior modeling efforts. Equations that show this achievement are provided in Chapter 8. Some details relating those properties to Einstein's special relativity and the concept of relativistic mass are discussed in the next chapter.

4. Special Relativity in the Hyperfluid

This chapter re-interprets Einstein's Special Relativity in terms of the hypervortex model. The hyperfluid becomes the causal factor that creates our observation of relativity. Einstein's results are used to determine some of the detailed properties of the hyperfluid and of the Hyperverse. The hypervortex model then explains, in simple physical terms, our observation of certain previously mysterious properties of particles.

Principle of Relativity: The principle of relativity predates Einstein's derivation of special relativity. Galileo stated the principle of relativity as, "it is impossible by mechanical means to say whether we move or stay at rest, so long as the motion is uniform and not fluctuating this way and that." Galileo developed his principle from observations of the motion of ships on water. He was noting that absent the ability to look out from the ship no experiment within the ship would give an answer that depended on the velocity of the ship. Even if the ship's motion is fluctuating one cannot tell whether the ship is anchored or achieving any overall net velocity. The fluctuations do allow one who knows the ocean to estimate, based on the size of the fluctuations, whether the ship may be in a protected harbor or more likely in an exposed area of the ocean. However, the fluctuations do not allow determination of the ships velocity, relative to the water, relative to the land or relative to the stars. The name of the concept, .i.e. "Principle of Relativity", derives from the implication that any local region of space, e.g., the ship, could be declared as being at rest (i.e. not moving) with all other objects assigned velocities relative to that rest frame.

Einstein applied Galileo's principle of relativity to interpret the results of experiments that used local measurements to measure the speed of light. In the experiments that measured the speed of light, the Earth is the ship from Galileo's definition of relativity. The experiments found the speed of light to be independent of Earth's motion around the sun. (Looking at stars in the distance is not a local measurement.) Einstein did not change Galileo's principle of relativity. Rather he noted that experiments had shown that measurements of the speed of light could not be used to determine our motion in the universe. Effectively, he noted that Galileo's claim that no local measurement inside the ship could be used to measure the velocity of the ship applied even to measurements of the speed of light and even to the circumstance when the Earth was the ship. Stated differently, Einstein's observation was that, if the measured speed of light had been dependent on the motion of the Earth, such a fact would violate Galileo's principle of relativity.

Leveraging his own observations, Einstein used the principle of relativity to derive properties of clocks, rulers and motion such that the principle of relativity would explain why all of the local measurements of the speed of light provided the same measured speed. Einstein provided no root cause for why relativity is obeyed. When he published his theory it was a derivation of the properties of a fluid through which light travelled; it was a derivation of properties of that fluid such that Galileo's principle of relativity would be upheld.

Einstein's early work on this topic became known as Special Relativity to distinguish it from his later work called General Relativity. Einstein's theory of special relativity applies to the special case of the principle of relativity when the condition applies, "… so long as the motion is uniform and not fluctuating this way and that". Einstein's theory of General Relativity refers to the inability, sometimes, to determine motion even when motion is non-uniform. In space, in orbit, people cannot measure velocity or acceleration by local measurement even while their spacecraft is in non-constant velocity as it orbits the Earth in free-fall. Einstein's work on General Relativity applies in such cases.

The principle of relativity applies in an ocean of hyperfluid as it does in ocean of water. Whether it applies for all possible hyperfluids is an interesting philosophical question, but not the point here. Rather, the point here is that, using Einstein's methods, we can derive at least a subset of possible hyperfluids for which the principle of relativity applies. Further, this has been achieved. At least one hyperfluid exists such that the principle of relativity applies, both in the special case and in the general case. Moreover, for that hyperfluid, no curvature of space is required. Also, no mixing of time and space is required.

Hyperfluid that Obeys the Principle of Relativity: A key property of a hyperfluid that obeys the principle of relativity is that there be a minimum of four spatial dimensions plus time; also, the spatial dimensions must have substantial if not infinite extent. For such hyperfluid, clock rates and ruler sizes can vary per Einstein's predictions. Einstein used the principle of relativity and the measured constancy of the speed of light to compute observable properties of a space-filling fluid. However, his mathematical solution, because it used only three spatial dimensions, required mixing space and time. Thus, his solution was unsatisfying when viewed as describing a space-filling 3D fluid, and consequently the more modern interpretation of space time curvature was invented. The hyperfluid makes Einstein's solution and his original interpretation satisfying and removes any need for space-time mixing or curvature.

Whereas Einstein noted that the measured speed of light must be independent of the speed of the measurement system in order that one cannot use variations in its measured value to determine the speed of the measurement system, the same must also be true for other properties of the universe. For example, if the measured mass of a particle (at rest relative to the measurement system) varied depending on the speed of that system, then measurements of the mass of such particles could be used to determine the speed of the measurement system, which would violate Galileo's principle of relativity. Similarly, if the measured charge or size of a particle at rest relative to the measurement system varied depending on the speed of that system, then measurement of those particle properties could be used to determine the speed of the measurement system.

In general if the measured value of any property of an object, measured from inside the object, changed as the speed of the object changed, that could be used to determine the speed of the object by local measurement. That would violate the principle of relativity. But Einstein said that the mass and length of objects does change as speed changes. He also said that clock rates change. The hypervortex model shows that many properties of objects change value as the object changes its speed relative to the ambient hyperfluid; the hypervortex model also shows they all change together such that the measured values do not change. For example, when the length of the object gets shorter, our ruler inside the object also gets shorter such that we measure the same length. All objects are made of hyperfluid, which is flexible, not rigid. Such construction supports these changes in properties.

While we cannot measure the changes to an object from the inside, we can measure the changes from the outside – from a distance such that the object being measured does not affect the ambient hyperfluid located at the place where our measurement is being made. Einstein's Special Relativity provides the values for key properties of objects as measured from such a location. Einstein derived the ratio between the clock rate outside the moving object and the clock rate inside the moving object. Similarly he derived ratios for the measurements of size and of mass. The hypervortex model is formulated to provide the same answers for these ratios as those provided by Einstein's model. Also, the hypervortex model is derived such that the principle of relativity is obeyed for all the basic properties of objects.

Note that it has not been determined that every possible property of the universe obeys relativity. It may be that the measured values of some property, or properties, of the universe vary such that the speed of the measuring system can be determined by local measurement. If so, then the principle of relativity may have an exception. That is okay; the principle of relativity is an assumption, not a proven fact. If it is violated for any particular measurable property of the universe, that violation can be used to refine our understanding of the properties of the hyperfluid. Meanwhile, properties of a hyperfluid have been identified such that the principle of relativity is obeyed.

The Laboratory Frame: The principle of relativity applies to local measurements. Local measurements require a local laboratory. The local laboratory is the place where the local measurement is made. The proper detailed definition of the local laboratory is a key aspect of determining the detailed properties of the hyperfluid such that measured results are constant when required and not necessarily otherwise. If a measurement involves space beyond the local laboratory, the measured results of a property of the universe need no longer be constant; relativity requires observation of constant values only when measurements are entirely local.

In Galileo's view of relativity as applying to a ship on an ocean, the ship is the laboratory. Further, for purposes of the application of the principle of relativity, the ship is a rigid structure, a uniform space, separate from the ocean. If a property of space is measured involving only space of the ship, then its result must be independent of the speed of the ship. Measurement of a star is not a local measurement. Measurement involving the distant shoreline is not a local measurement.

In the hypervortex model, the laboratory is composed of vortices of hyperfluid, and ergo the laboratory is made of hyperfluid. Because of this, the laboratory is not rigid like a ship but rather flexes following every local change in the flow of the expansion current. This flexing behavior has been a factor contributing to the difficulty of realizing how Einstein's special relativity can occur in a fluid. The impact of this non-rigidity is difficult to describe fully; it is perhaps sufficient to think of two different laboratory frames at two different size scales and to understand the relation and transition between them.

At the small scale, consider a laboratory frame that encompasses a single particle and some free space around it. In these particle reference frames, in order that relativity is obeyed, the axis of the particle's vortex is always parallel to the local expansion current, and perpendicular to the three observable axes of the laboratory. If the orientation of these elements was not so, then a measurement of that orientation could be used to determine the velocity of the particle, violating relativity. Also, in that frame, clock rates change and rulers change so that the measured value of all major properties of the particle are constant, no matter what its speed and relativistic mass might be in other reference frames.

At the larger scale, consider a laboratory frame that encompasses the scope of a real experiment of interest. This experiment frame will have a constant shape as it flows with the expansion current, if and only if its span is limited to a region of space across which the expansion current is uniform, except as it may be distorted very near individual vortices in the laboratory. This laboratory is always at rest with respect to the bulk hyperfluid in the region of space encompassed by the laboratory; it corresponds to a center-of-mass reference frame as defined in standard physics. The orientation of the experiment frame's axes varies relative to each particle's reference frame and the relative orientation specifically depends on the velocity of each particle. The specifics of the relationship between the axes of the experiment frame versus that axes of each particle frame create the observation of Einstein's special relativity.

The clocks and rulers in the frame that is at rest relative to the local expansion current are the correct ones for computing various effects, especially relativistic effects that may occur in any experiment. Mathematically, one can do the calculations in a wide variety of reference frames. In some cases, calculations done in any reference frame will be correct. However, many times the actual velocity of the hyperfluid and the expansion current affect the result. In such cases, the mathematics of physics, both in standard models and in the hypervortex model is written such that the calculation done in the frame in which the hyperfluid is at rest transparently gives the correct results. Use of any other frame requires complex transforms to account for the velocity of the hyperfluid.

Some experiments do extend over a region of space such that the expansion current varies over space and even over time. Such experiments can be modeled in the hyperfluid by properly accounting for the variations in the hyperfluid flow. That is discussed in the next section, in the context of the example of the twin paradox, a standard paradox of special relativity.

Variable Clock Rates: In his theory of special relativity, Einstein computed that clock rates change for particles in motion relative to a given laboratory. He showed that these changes were required in order that the principle of relativity be obeyed. In particular, he showed that the clock rate of a particle moving relative to a given laboratory frame slows down relative to a clock at rest. This has been confirmed experimentally many times. Einstein's theory, however, did not explain why the clock rates changed. As a result there have been reasons for this phenomenon whose debate borders on the metaphysical. At one extreme, Option A, the clock rates change simply because clocks do not directly measure time, rather they simply repeat a motion at a rate that depends on local factors. In such interpretation there is a global time, but we must adjust our interpretation of the local clock to know the current global time. At the other extreme, Option B, there is no global time; time is a complex space-time coordinate and space curves and deforms around each particle and object which changes the local time hence the local clock rate.

Particle models have adopted the latter extreme view of time, Option B, but using it, the models have been unable to complete their goal. Meanwhile engineers have adopted Option A and, using it, clocks around the world on satellites and across the solar system are unified to a global time that works.

The hypervortex model provides a unifying view of these two extreme views of time. In the Option B view of the world, the rate of clocks can be viewed as depending on an angle that depends on the observed velocity of the particle. However, in the standard mathematics there is no physical meaning to such an angle. In the hypervortex model that angle is the angle that the vortex makes with the expansion current. When a vortex is aligned with the overall expansion current of the laboratory frame in which is resides, the observed velocity of the corresponding particle is zero, and local clock rate of the vortex in its local particle frame matches the clock rate in the laboratory frame. As the observed particle velocity changes that angle changes in just the way that Einstein's equations prescribe; the hypervortex model says that the vortex deforms to become longer and larger in all dimensions and that the local clock rate in the local frame of the vortex slows down correspondingly. Thus in the local frame, all measured values of the particle are unchanged. Meanwhile that clock rate is now slower than the laboratory clock rate, and by the amount computed by Einstein's equations.

Velocity is not the only property that can change a clock rate. Changes in the velocity of the expansion current and changes in the density of the hyperfluid can also change the clock rate. All this still appears to occur in accord with Einstein's equations. The difference is that whereas standard interpretations of the equations declare that the fabric of space is curved without the provision of any physical cause, in the hypervortex model, the observable universe is a curved changing slice through the hyperfluid due to these deformations of the hyperfluid that change clock rates.

The Twin Paradox: The twin paradox is a long standing scenario that was created to challenge and explain various interpretations of variable clock rates. The standard scenario starts with twins initially at some location. One twin then goes off at relativistic speeds and later turns around and returns. A simple view that uses the standard interpretation creates the paradox. It says that each twin, using his own local laboratory frame as the basis for his calculations, calculates that the other twin's clock slows down and so each says the other twin is younger.

The standard resolution to the issue was debated for decades. It is complicated and hopefully about to become obsolete, so it will not be discussed in detail here. Importantly, the explanation that is about to become obsolete requires giving up the idea of absolute simultaneity – the idea that two things either did or did not occur at the same time – which in turn requires giving up the idea of any sense of a global time. The explanation also requires giving up any possibility of something traveling much faster than light, since such a thing would restore absolute simultaneity and void the current explanation of the twin paradox.

The hypervortex model provides a straightforward non-paradoxical answer. The twin ages slower, whose corresponding vortex is at the larger angle relative to the overall direction of the expansion current for more of the time. In the twin paradox scenario, the correct calculation is made by the twin who is at rest relative to the laboratory frame of the hyperfluid in the region of space where the twins travel. The other twin's calculation is incorrect. The clock and reference frame that he uses for his calculation, while valid within his ship is not valid across the space between the twins. That twin is assuming that he is at rest relative to the hyperfluid and he is not.

There are big differences between the standard explanation of the twin paradox and the hyperfluid explanation. Both explanations use the principle of relativity; however the standard explanation of the twin paradox applies the principle beyond its bounds of validity. Recall that Galileo stated the principle of relativity as regards inability to know the speed of a ship by local measurement. The local measurement part of his statement is critical. His statement implicitly also states that in order to know the actual speed of the ship one must know the state of the water outside the ship. This latter statement is called a corollary to the stated principle. The twins in the standard twin paradox ignore this corollary; they ignore everything outside their respective ships when making the calculation of the other twin's age. They both pretend that the corollary does not exist, that from inside their ship they can know the actual velocity of their ship and of their twin's ship.

In the scenario stated for the twin paradox, one twin happens to be at rest relative to the hyperfluid so his calculation happens to be correct. If the motion of the hyperfluid between the twins is more complicated across the space travelled by the traveling twin, as it likely will be whenever this experiment can be performed, knowledge of the state of the hyperfluid across space will be needed in order to compute a result that matches the measurement.

Observing a Constant Speed of Light: It is "well known" that, "The speed of light in vacuum is a constant." There is more to it. The speed of light is independent of the speed of its emitter. It is also independent of the speed of the laboratory frame in which it is measured. These facts were measured, and Einstein developed the theory of special relativity to compute the properties of a fluid such that this would be true.

The statements made about the speed of light apply for all waves in all known real fluids. All waves in all known fluids travel at a characteristic speed relative to the fluid in which it travels. In any real laboratory frame, where the fluid is at rest relative to the laboratory, the measured speed of the wave will be the same characteristic speed. When sound is traveling inside a train it is traveling at the speed of sound relative to the air in the train. When the sound is emitted from the train it then travels at the speed of sound relative to the air outside the train and hence relative to the ground, unless the wind is blowing. The speed of the train when it emits the sound is irrelevant to the speed of the sound after it leaves the train. The same is true with waves in water. The speed of the boat's wake is independent of the speed of the boat. The speed at which whale songs travel is independent of the speed of the whale. The same is true in super-fluids such as liquid Helium 3. The speed of waves in fluids is relative to the speed of the fluid and it can depend on the density of the fluid and other properties of it. For a region in which the fluid properties are uniform, the speed of the wave relative to the rest frame, or laboratory frame, is constant.

There was a time when light was predicted to move in a special fluid whose properties would be an exception to other fluids. In that special fluid, the measured speed of waves was predicted to depend on the speed of the laboratory. The idea of this exceptional fluid was developed due to other physics, not due to properties of light. This was discussed some in Chapter 1 and will be discussed more in Chapter 6. Measurements proved that particular prediction wrong by showing that the speed of light is measured to be the same in all laboratory frames. Einstein developed equations that avoided the need for this exceptional fluid. He found that, whether or not an exceptional fluid existed, variable clock rates and a mixing of space and time could explain the relevant multiple aspects of physics.

The hypervortex model avoids assigning the fluid exceptional properties. To date, an ideal hyperfluid with rather ordinary properties is sufficient to support the observed properties related to relativity – clock rates that depend on their motion, gravity and other factors. Just as the speed of sound depends on the density and temperature of air, and other factors, the speed of light can depend on certain physical properties of the fluid, but not the speed of the laboratory.

An observant reader might note a particularly interesting question and challenge the observation of a constant speed of light when clocks have varying rates. The basic answer is that the speed of light is measured by clocking the time required for light to travel a given distance. When a particle or object is moving through a laboratory so that its clock slows down, in turn, the particle also physically expands in all directions and by the same scale. It was discussed earlier that accelerating a vortex adds energy that is used to lengthen the vortex. That is part of the expansion. The vortex and region around it expands in all dimensions so that any measurement scales to give the same answer no matter the speed of the clock local to the vortex. The laboratory can observe some changes to the particle as it expands, such as its increasing mass, but it does not see a change in the speed of light at that particle, since the speed of light is still the same.

The requirement that the speed of light is measured as constant in all reference frames no matter the speed of the local clock is one key reason that this model is of a hyperfluid rather than of a three dimensional fluid. Some of the ideas presented here are part of the historical efforts. However, when tried with three dimensional fluids no model was found consistent with all of the necessary aspects. The details of those prior efforts will not be presented here. Many scientists attempted such fluid models when Einstein's results were first proven right. However, without the additional spatial coordinate used here to create a hyperfluid, time must serve as both a spatial coordinate and a time coordinate. The requirement for time to perform this dual function over-constrained those models. Splitting the two functions into separate factors, a fourth spatial dimension and a separate time coordinate enables success of the hypervortex model.

The Observed Particle Mass: The source of particle mass was introduced in the discussion of mass energy equivalence. There is much more to particle mass than was introduced there. To recap first, the observed mass of particles increases as their speed in a given laboratory increases. This was predicted by Einstein and measured.

His equations did not explain it. The hypervortex model explains that the increase in mass is the observable effect of the increase in the particle size as the clock rate slows and the vortices expand. As the objects accelerate, they grow in all four spatial dimensions. This growth requires energy, and it is the reason that the acceleration of any object requires energy. The physical growth in the observable spatial dimensions does not require energy, but the growth along the fourth spatial dimension does require energy. Remembering that each particle is the observable portion of a vortex, the growth of the particle along the unseen spatial dimension is really an increase in the length of the vortex. The vortex has a fixed mass per unit length. Its observed mass increases directly as its length of vortex inside the observable universe grows, which occurs because of the change in orientation of its vortex relative to the expansion current and hence to the observable universe. This mass growth is the observable evidence that the particle size grows.

The changing size of a vortex as the local properties of a fluid change has analogy in other fluids with other eddy currents, other vortices. One good example is tornadoes. Near the ground where the air is densest, the diameter of the tornado is smallest. At the top of the tornado, where the air is thinnest its diameter is the largest. The size variation is similar for eddy currents in water. In particular, for eddy currents that start at the surface of the water and penetrate into the body of the water, the diameter of the vortex shrinks as the distance from the air/water boundary increases. Meanwhile, the diameter of vortices can be constant when generated where the properties of the fluid are constant. For example, a vortex can be observed entirely under water by creating appropriate turbulence with a sweeping motion of the hand underwater. The water conditions are constant along the length of the vortex, so its diameter is constant. Vortices in the hyperfluid are similar. Motion of the vortex relative to the hyperfluid changes the hyperfluid properties so that the vortex changes size.

Einstein noted that, in principle, particles could have two different observable masses – an inertial mass that affects the rate at which they accelerate when a force is applied to them, and a gravitational mass that determines the gravitational force that a particle generates. He further noted, or postulated, that these two masses are the same, or at least observed as if the same. He did not explain the equality of these two masses. Moreover, no particle model can explain the equality of these two masses – one is quantum effect and the other is a gravitational effect and no particle model has unified quantum mechanics and gravity.

One substantial accomplishment of the hypervortex model is that its properties have been determined such that the gravitational and inertial masses of any object are the same. Both masses are directly proportional to the energy per unit length of the vortex that we observe as a particle. The radius of the vortex is proportional to observed mass of the vortex and the radius of the vortex is proportional to the observed gravitational force that it generates. Also, the energy per unit length of the vortex is proportional to the force required to accelerate the particle (because the force actually is applied to increase the length of the vortex and thus is proportional to the inertial mass of the particle). Thus, the inertial mass and the gravitational mass are proportional to each other because they are both proportional to the characteristic size of the particle.

Note that the fact that both the inertial mass and the gravitational mass increase as a particle's velocity increases is evidence that the vortex size scales equally in all spatial dimensions as its velocity increases. The increase in inertial mass is evidence that the vortex gets longer and changes orientation as its speed increases. The increase in gravitational force proportional to the increase in mass is evidence that the particle expands in radius.

Relativistic Particle Motion: A key result of Einstein's special relativity is that it limits the observed speed of particles to be less than the speed of light. Standard models expand the scope of this limit to include everything, while the hypervortex model shows clearly that the limit applies to particles, light, and perhaps some things but not everything.

The reason that the limit occurs is in part that, when a particle is accelerated towards the speed of light the amount of energy required to further accelerate becomes infinite. However, the speed limit is also due in part to how the expansion current interacts with the particle's vortex to create the observed motion. The particle does not really move just because we push or pull it. Rather, our actions change the vortex's orientation relative to the expansion current as described in prior sections. The expansion current pushing on the vortex then moves it, which creates our observation of particle motion.

Consider the analogy of the wind pushing on a kite. The string and kite are like a vortex being blown by the expansion current. More specifically, think of the kite as the particle that we observe and the string as the remainder of the vortex that we cannot observe. We can change the tilt of the kite relative to the wind which changes the motion of the kite in the direction perpendicular to the wind. It similarly changes the motion of the string. Similarly with another analogy, a sailboat; we set the orientation of the boat and sail relative to the wind and the wind moves the boat. The kite's speed is directly related to the wind speed. The boats speed is related to the speed of the wind and to the speed of the wake that the moving boat creates. The boat has two limiting factors because it is at the surface between two fluids – air and water. The specific relation between the wind speed and the kite or boat's speeds depends on details. In the case of the hyperfluid, the expansion current flowing past a vortex creates the speed limit on the particle that matches the predictions of Einstein's special relativity.

While we observe the particle's motion in three dimensions, at the intersection of the observable universe and the vortex, the vortex's motions are more complicated and four dimensional. The motion of particle vortices is such that they satisfy two factors. First, vortices move such that the observed speed of the observed particle and the orientation of the vortex relative to the expansion current are quantitatively related to satisfy Einstein's special relativity. Second, particle vortices move through the observable universe at the speed of light. Thus, at each moment a different piece of each particle's vortex is intersecting any particular observable universe.

The second of the above factors, derives from the variable rates of clocks. The first method that was used to prove that clock rates slow down for particles moving at relativistic speed was to time the decay of short lived particles. Since particles will be observed as decayed when their vortex no longer intersects the observable universe, the rate at which the vortex passes through the observed universe must be such that the time for which a vortex intersects the observable universe matches the lifetime predicted by the particle's clock, at all orientations of the vortex relative to the expansion current. This requirement is satisfied if the vortex passes through the local observable universe at the speed of light.

The four dimensional motion of vortices, in the absence of forces, satisfies the two above factors. For a particle observed at rest, the corresponding vortex travels at the speed of light along the direction of the fourth spatial dimension. The vortex moves in the direction opposite that of the expansion current. As the observed particle velocity rises, the speed of the vortex declines along the fourth dimension, as measured in the, laboratory frame. As the observed particle velocity becomes relativistic, that speed drops significantly and continues to drop towards zero as the observed velocity rises towards the speed of light. The equations that quantify the speed as it depends on the orientation of the vortex to the expansion current are given by the appropriate factors provided in Einstein's special relativity.

The speed of light is the same as the speed of the expansion current. This is not a coincidence. The expansion current pushes the waves that we observe as light the same way that it pushes all vortices; it pushes light waves at the limiting speed. However, light is unique in that photons of light are emitted perpendicular to the vortices of particles. Thus, in the absence of forces, the photons usually have no motion in the direction perpendicular to the observable universe. Thus, each photon usually stays within the same observable universe over time. The photons follow curved paths so that they travel perpendicular to the expansion current and perpendicular to every vortex. Their motion defines the observable universe.

Curvature and Coordinate Systems: Deformation of the shape of the observable universe, caused by non-uniformity of the hyperfluid motion near each vortex causes a particle configuration that creates curvature of the observable universe. At the very location where the vortex intersects the observable universe, the vortex is always perpendicular to the 3D observable universe. That occurs because the observable universe distorts its thickness and orientation at that location to maintain perpendicularity to the vortex axis. A very short distance away, still within the laboratory rest frame, the angle between the vortex and that rest frame varies dependent on the particle's observed velocity. For a particle observed at rest, the vortex and laboratory frame are perpendicular. As the observed particle speed rises the angle becomes smaller, dropping to zero in the limit of particles traveling at the speed of light.

Coordinate systems are a means to reference every point in space by a unique set of numbers. Einstein's coordinate system used for general relativity is designed to curve to follow the shape of one observable 3D universe. It can thus reference only events occurring at locations in that one observable 3D universe. Since photons are the objects that stay within one observable 3D universe, Einstein's coordinate system in fact curves to enable referencing only the points in space where a given photon might travel. Einstein's coordinate system curves because hyperfluid distortions cause light to curve. Light does not curve because space curves; light curves because the expansion current curves.

The transition to the hypervortex model with its four spatial coordinate systems and time is akin to the historical transition from representing locations on Earth using curved coordinates – longitude, latitude and altitude – with Earth at the center to a coordinate system using the Sun as the center. Using Earth-centric coordinates has often been convenient, and often still is. However, a view with Earth stationary at the center of the universe as represented by such coordinates prevented our understanding of many phenomena. To understand things like the formation of hurricanes and ocean currents, a coordinate system was needed in which the Earth is a rotating body in motion. It is the Earth's rotation that causes the hurricanes and ocean currents. Thus, study of the Earth as a moving body in a larger space using flat rather than curved coordinates has been much superior for such study. Similarly, transition to the hypervortex model and its coordinate system is much superior for study of the large scale behaviors of galaxies and the observable universe itself in the larger Hyperverse.

Rigorous theorems prove that any coordinate system in which the motions of all of the objects in space can be represented can be used to study physics. Thus, it is absolutely possible to use four spatial coordinates and time instead of three curves spatial coordinates and a mixed space-time. The curved spatial coordinates with a mixed space-time reference a changing subset of locations in the larger flat space. Transform between the two coordinate systems involves the rotation and scaling that have been qualitatively described in prior sections. The rotation and scaling depend on the motion of the hyperfluid. Understanding the outcome of the transformation requires understanding that the particle model's coordinate system, with fewer dimensions references only the projection of hyper-objects onto the observable universe.

The hypervortex model uses the larger flat coordinate system to compute and understand physical phenomena and the full behavior of hyper-objects. It then uses the transformation to relate the real physics to our limited observation of it. In contrast, particle models use curved space-time coordinates directly and therefore cannot discuss reasons for the curvature or the effects of other parts of space. Also, understanding the mathematics and the physics therefore becomes conceptually more complicated in the particle models.

Energy Conservation in Each Universe: The energy in each observable universe is observed as being conserved. Particle models do not explain this; they simply assume it to be true. In contrast the hypervortex model provides explanation and detailed physical foundation in terms of the laws of motion and the properties of the hyperfluid. Generally speaking, forces between vortices occur along a line perpendicular to them. Also, light generally travels in directions perpendicular to the vortices. Further, each 3D universe is locally perpendicular to the vortices. Thus, force lines, which exchange energy between vortices, generally lie within a 3D universe. They thus generally exchange energy between parts of vortices within the same 3D universe. Therefore, energy and information exchange generally stay within 3D universes within the larger Hyperverse. This is why the Hyperverse divides into 3D universes. There is no physical structure that divides the Hyperverse into 3D universes. There is no physical boundary between different 3D universes. Rather, it is simply that, because of the way forces work and light moves, perpendicular to the expansion current of the Hyperverse, there exist subsets of locations in the Hyperverse such that observers within that subset can sense and exchange information with, and usually only with, other locations in that subset. We call those subsets 3D universes, observable universes, or simply universes.

Certain exceptions exist to exchange energy and information between observable universes. While most light travels within a 3D observable universe some may scatter or diffract into a neighboring universe. Also, information can travel along vortices at times. Recent experiments have seen such effects, although most models do not understand their significance as inter-universe exchange, and thus have real difficulty explaining the phenomena. Discussion of such physics comprises a subject called quantum mechanics.

5. Quantum Mechanics in the Hyperfluid

The nature of quantum mechanics and its relationship to other physics is perhaps best introduced using an analogy from the previous chapter. In the previous chapter an analogy was established between a kite in the wind and a vortex pushed around by the expansion current. The analogy was introduced to discuss how the overall motion of a vortex is akin to the overall motion of a kite on a string – that our actions on the string control what the wind does to the kite, but the wind is the actual cause of the kite's motion (and the expansion current is the cause of the particle's motion.) Reusing that analogy here, quantum mechanics essentially discusses details of the analogy that were not discussed in the previous chapter. For example, in addition to moving around in the wind, the kite and the string vibrate due to the wind; the kite and the string do not vibrate when there is no wind. Also, there is a probability that the kite and the string will break in the wind. Also, there is a possibility that the kite and string will become entangled with another kite and string. Quantum mechanics is truly about all these details as they apply not to the kite string, but rather to vortices in the hyperfluid.

To start applying the kite analogy to quantum mechanics, consider five facts: (1) sound in air travels much faster than wind, (2) sound traveling along a string is a vibration of the string (3), sound travels along strings much faster than in air, (4) the expansion current in the hyperfluid is analogous to the wind on the kite, and (5) the speed of light is the same as the speed of the expansion current. Combining these five facts, the analogy suggests that just as the speed of sound on the string is much faster than the speed of the wind, hence, the speed of vibrations along the vortex is much faster than the speed of light. Experiments have observed a faster than light phenomenon in quantum mechanics. Those experiments confirmed Einstein's predicted "Spooky Action at a Distance". In standard models faster than light travel is a contraction to the foundations of relativity theory. In contrast, in the hyperfluid there is no contradiction; the speed of light does not limit the speed of vibrations on a vortex just as the speed of the wind does not limit the speed of vibrations on a string.

Sections of this chapter describe various aspects of quantum mechanics using further analogy to the kite and the string, or sometimes the analogy to a boat and its wake, which analogy was introduced in a previous chapter.

Heisenberg Uncertainty: One key aspect of quantum mechanics is its finding that our ability to know the physical location and velocity of a particle includes a fundamental uncertainty. In the hypervortex model, this uncertainty is a result of the vibrations of the vortex along its length. Just as in the analogy to the kite and the string, vibrations along the kite and string cause uncertainty in our knowledge of the instantaneous location of the kite and the string, the same is true of vortices. We know on average where the kite and string are and where they are moving. However, at any given moment we do not know exactly where they are or where they are moving. In principle the exact location of the string and kite is knowable at every instant; in practice vibrations limit our precise knowledge. Rather, we know that they are within some relatively small distance of their average location. We further know that the details of how far they can move from the average location are complicated and determined by the vibrations and the properties of the string and kite.

The reason that vortices can support vibrations is that their length exceeds the length necessary to follow the path predicted for it by classical mechanics. Classical mechanics used to be called just "mechanics", until quantum mechanical phenomena were discovered. Then "mechanics" became named "classical mechanics" to distinguish it from quantum mechanics. Classical mechanics computes the average location and velocity of the vortex; it computes where the vortex would be in space and time if the length of the vortex was the minimum length required to follow the path computed for it by classical mechanics. Quantum mechanics notes that the length of the vortex is everywhere longer than that minimum length. Just as the typical kite string includes twists or braiding in its structure that lets the string easily stretch, bend, flex and vibrate, so too does the vortex. The extra length lets the vortex move outside the path computed for it by classical mechanics.

Heisenberg quantified the amount that a vortex's path could deviate from the classical path. He did not know that he was doing so; he wrote his equation in terms of a particle. It is the hypervortex model that tells us that his equation is equivalent to a statement quantifying the surplus length of vortices. Again the analogy to a string helps explain the science. A string with some surplus length can have many small vibrations of short wavelength or fewer larger vibrations at longer wavelengths. The amount of surplus length determines the relation between the amplitude and the wavelength of the vibrations. For vortices in the hyperfluid, Heisenberg quantified this as $\Delta x \Delta p \geq h/4\pi$, where Δx has been assumed in typical past interpretations to be the uncertainty in location of a particle, Δp has been interpreted as the uncertainty in its corresponding component of momentum, and h is Plank's Constant. The equality has been assumed to apply when measurements introduce no additional uncertainty; otherwise the inequality applies.

The hyperfluid changes the interpretation of Heisenberg's Uncertainty principle in three ways. First, the vortex lies along a certain, not uncertain, path. That is, at every moment every part of the vortex has a specific velocity and location. Second, Δx is the deviation of that path from the classical path. Third, because of how the particle momentum relates to the angle between the vortex and the expansion current, Δp captures the rate of change, along the vortex, at a snapshot in time, of the deviation of the vortex path from the classical path. Further, the deviations have periodicity along the length of the vortex and Δp is inversely related to the wavelength of that periodicity.

Despite the fact that the vortex has a specific location and motion at every point along its length at every moment, we can only measure the vortex location and velocity in the one observable universe in which we reside. From that measurement we cannot compute the exact motion of the vortex, or even of the observable particle. Note that the problem is a fundamental lack of information. We, as 3D observers, cannot predict the future motion of a particle from such a measurement no matter whether our measurement of the particle changes its behavior. Thus, uncertainty of particle locations and their motion over time persists as an observable fact of our existence.

Stable and Unstable Particles: Stability of a particle, a vortex, or anything, refers to its ability to retain its same state. In the case of a particle or vortex this refers to its physical state. Understanding certain details about stable and unstable particles helps our understanding of the vibrations along the length of vortices, which improves our understanding of quantum mechanics.

Some particles are very stable. Other particles are metastable or even unstable. Metastable particles tend to remain unchanged until some trigger, even a minor one, induces a change. Unstable particles change even when no external trigger occurs. The same particle can be stable, metastable or unstable depending on its circumstances. This is true for fundamental particles such as electrons and for composite particles such as atoms. Much of the early research in quantum mechanics examined atoms changing from unstable states into stable states. The fact that these changes occurred via the emission of quantized amounts of light waves (i.e. photons) is a reason that the study is called quantum mechanics.

Many unstable particles become stable by changing into wholly different forms (e.g., into multiple particles and photons). Other unstable particles emit energy to become a less energetic version of the same type of particle. Both types of instabilities change by emission of quanta of energy, usually light; both are important to understanding quantum mechanics. For all of these unstable particles, the length of the corresponding vortex (or cluster of vortices) directly relates to the length of time for which the particle exists. A corollary of Heisenberg's uncertainty principle, expressed in terms of the hypervortex model, is that the variability in the length of vortices is inversely related to the lifetime of the particle; the shorter it lives the more variability there is in the length of its corresponding vortex or vortices. Similarly, the longer a particle lives, the more consistent the length of the vortex or vortices. Actually, Heisenberg's principles states this in terms of a relation between the particles lifetime and its energy; the hypervortex model shows that the energy and length of the vortex are directly related, which enables us here to discuss the relation between particle stability and vortex length.

Stable particles are our observations of vortices that are long lived at constant energy. Electrons and protons have lifetimes discussed in terms of billions of years. Thus, the energy, the mass, and the vortex length of all electrons are the essentially the same and constant. Similarly, the energy, mass and vortex length for protons are essentially the same for all protons. Thus, vibrations along the length of these particles do not change their length. Such a constraint limits the types of vibrations that are sustainable along stable vortices. In particular, it suggests that the preferred vibration modes are helically structured. Think of a coil spring, or the twists in a kite string, or best perhaps, the phone cord connecting the handset to the base of a classic wire-line telephone.

The classic helical telephone cord displays exactly the functional properties needed for vortex vibrations. The observed length of the coiled cord can change. Also, it flexes easily and it supports vibration modes. Yet, the length of the actual wire inside the cord does not change, nor is there even any substantial stress on it – similarly for a vortex configured into a helical cord. To be clear, the dimensions of the helical vortex are different than for the phone cord. Also, the dimensions of the helical features can vary. Also, to be clear the helical superstructure is quite distinct from the axis of the vortex and the remainder of the vortex structure. The radius of the helical structure may be a million times larger than the radius of the central void of the vortex, or more. If the central void of an electron were the size of a small tornado, the radius of the helical structure could exceed the radius of the Earth.

When a vortex is configured to form a helical superstructure, the vortex essentially orbits around the classically computed path for the vortex. The radius of the helix corresponds to the quantum mechanical uncertainty of the location of the particle. The period of the helix directly relates to quantum mechanical uncertainty in the momentum of the particle. The period of the helix is the distance, along the length of the vortex required for the helix to make one orbit around the classical path. The momentum of the particle relates to the angle of the vortex relative to the expansion current, and as the vortex orbits around the classical path, that angle varies thus creating the uncertainty of the particle's momentum.

Quantization: Quantization is the aspect of quantum mechanics from which the specialty area of physics derives its name. Early research in quantum mechanics found that when an atom spontaneously changes from an unstable state to a stable state it emits a discrete quantized unit of light. Similarly any particle changing from any state to another state emits or absorbs a quantum unit of energy. The state changes do not absorb arbitrary amounts of energy but rather discrete specific amount of energy.

In the hyperfluid, quantization in the hyperfluid directly results from the fact that certain properties of ideal fluids can cause them to support only specific stable structures that contain discrete and well defined amounts of energy. All other structures in the hyperfluid are meta-stable or unstable. The discrete nature of the energy spectrum is observable for structures in the stable state for that structure type. It is also visible for states that are relatively close to the stable state. This is because the difference between the energy in the stable state and the energy in the other states is a relative fraction of the energy content of the stable state. We call this regime of observation quantum mechanics. When we observe higher energy states, the energy differences are small compared to the total energy and so are hard or impossible to measure. We call that regime classical mechanics. The detailed properties of the hyperfluid determine the set of observable states. Thus, measurement of the states can be used to determine some detailed properties of the hyperfluid.

Quantization in the hyperfluid occurs at two levels. One level regards the structure and stability of fundamental particles such as electrons and protons. The other regards combinations of such particles into atoms (and molecules). Initially quantum mechanics was about the latter. However, over time, as part of the effort to unify all physics, quantization was also found to affect the fundamental particles. Particle models deal with both, as does the hypervortex model. The differences are substantial.

In the hyperfluid view stability occurs in a fundamental particle when the various forces within the corresponding vortices exist in a stable balance. The radial variation in the fluid density creates a force that tries to move fluid to fill the void at the center of the vortex. Meanwhile, the rotation of the fluid around the central void creates counter forces. These counter forces include a centrifugal force and forces due to variations in hyperfluid velocity. When all of the forces are in balance the particle structure is stable. The details of this balance are different for different types of vortices. Leptons (including electrons, muons, and neutrinos), mesons, and baryons (nucleons – including protons and neutrons) each manifest a different type of vortex structure. Within those types are subtypes with different masses and other different details. Some sub-types are stable while others rapidly decay, often into the lowest mass member of that type. Some decay because they are inherently unstable, others because small interactions make them unstable. Even neutrons decay when not proximate to a proton.

The interaction between the basic vortex structure and the helical super-structure has a strong effect on the stability of atoms. In the early days of quantum mechanics, before it became a mostly mathematical endeavor, scientists noted that preferred energy levels of hydrogen atoms occurred when the length of the classically computed orbital path of the electron was an integer number of periods or cycles of a particular quantum mechanical wave that seemed to be associated with the electron. However, no real physical meaning was found for that quantum mechanical wave at the time. Also, the effort to compute the energy levels of more complex atoms by such method proved too difficult and its results too approximate.

The hypervortex model gives new meaning to the original physical view of the cause of the existence of preferred quantum energy levels. The original physical view of the quantized energy levels in atoms, the Bohr model of the atom, was that electrons preferred to be in orbits whose periodicity had certain properties. Electrons were observed to prefer orbits at certain radii and with certain periodicity. The preferences were observed to depend on the number of protons in the nucleus of the atom. In the hypervortex model, electrons orbiting atomic nuclei are our observation of vortices that have periodic structure along the length of the vortex. If the electron is in circular orbit around an atom then the corresponding vortex has a helical shape. That helically shaped vortex orbits around the set of vortices that corresponds to all the protons and neutrons in the nucleus of the atom. The preferred quantum energy levels of electrons in such orbits are those that use all of the available length of the electron's vortex. For other orbits, residual length of the electron's vortex enables additional vibration patterns on the vortex. Those additional vibration patterns allow instabilities in the vortex shape that tend to radiate energy until the vortex length is reduced to match a preferred energy level. This new understanding does not change the mathematics of quantum mechanics; it explains the physical meaning of the mathematics. It also helps us determine the degree to which the mathematics of quantum mechanics is an approximation to a more complex mathematics.

Least "Action": The principle of least action is a central principle to classical physics. "Action" is a specific standard physics quantity (not a new concept for this model) that describes the exchange of kinetic and potential energy. The principle of Least Action generalizes Newton's Law's to say approximately that, not only does a particle in motion tend to stay in motion, also, when it must change its motion it does so in a way that minimizes the overall sloshing of energy over time between kinetic energy and potential energy. The concept is central to classical physics and to the construction of the mathematics of many attempts to develop unified theories of everything. In Chapter 8, it will be used to develop the mathematics of the hypervortex model.

The reason that the concept of "least action" is relevant to a discussion of quantum mechanics in the hyperfluid is that in the hypervortex model, quantum mechanics fits naturally into the principle of least action. Each vortex traces a path through the hyperfluid. The location and motion of that path over time varies around the path computed classically by the principle of "Least Action". The classical path of the vortex is the path that it can trace with least action. The quantum path is a path allowed by generalization of the mathematics of least action to include paths similar to the path of least action. The specifics and full mathematical form to compute the exact quantum path as it deviates from the classical path is complex and yet compatible with current quantum theory. It is discussed in some detail in Chapter 8. It is also important to note here and in Chapter 8, that the mathematics and its relevance to quantum theory were well developed decades ago by others without regard to any particular model of the universe. The hypervortex model gives that prior effort the significance and relevance it has long deserved.

"Least Action" presents an important difference between particle models and the hyperfluid; in the hypervortex model the calculation of quantum mechanics is a natural extension of the classical computation. In the hypervortex model no distinct theory is required. In the hypervortex model, the full 4D structure of atoms is fully computable. However, we observe only a 3D subset at any given time. Thus, we lack the data required to do that computation and we lack the ability to see the underlying causes of certain phenomena. Also, with both the vortices and our 3D observable universe moving continuously in the 4D Hyperverse, the computations become uncertain. The following sections discuss each of these various aspects of quantum mechanics and its underlying cause in the hyperfluid.

Probabilistic Events: The above discussion has shown that quantum mechanics in the hyperfluid is about the ability of particles to deviate from the classically computed path because of the available extra length of vortex. As described above, the location and motion of the vortices in four dimensions is fully determined at every moment. Our observation of the probabilistic behavior of particles results from a combination of factors: (1) surplus vortex length enabling somewhat circuitous paths that are hard to predict, (2) continuous flow of fluid and vortices through our observable universe, (3) the existence of waves and vibrations along the vortex that affect its motion and (4) our inability to observe the full length of each vortex. The combination of these factors causes the location and momentum of particles to be observed as being statistically distributed. The statistical distribution is always centered on the classically computed location and momentum.

Einstein is quoted as saying that god does not play dice with the world. By that it is meant that he believed that quantum mechanics is not fundamentally probabilistic. The hyperfluid view agrees with his view and indicates that he may have been right. A being who can see the whole 4D Hyperverse and who can compute the kinematics of the vortex waves, can compute the exact particle motion at quantum time scales, including all deviations from the classical path. Meanwhile, lacking the required omnipotence, man appears stuck with a probabilistic approach to computing motion at the quantum regime.

Other models tried for decades to remove the probabilistic interpretation of quantum mechanics. Many sought to find a "hidden variable", the knowledge of which would remove the probabilistic nature associated with quantum theory. Others tried to prove that no 'hidden variable" could remove the probabilistic nature of quantum theory. The hypervortex model now provides insight regarding those efforts. In particular, the hyperfluid shows that what was missing was not a single variable but rather a whole spatial dimension. In order to remove the probabilistic nature of our observation of quantum events, we need to be able to simultaneously observe all of the nearby observable universes as well as our own. Until we can discover a way to observe four dimensionally, quantum mechanics will be inherently probabilistic.

Double Slit Experiments: "Double slit experiments" refers to a whole style of experiments developed and performed historically to study the nature of particles and light waves. Double slit experiments have been performed with variety of types of particles to try to understand probabilistic outcomes in the quantum regime. The outcome of those experiments is discussed here because of their great value towards illuminating the nature of vortices in the hyperfluid and the true nature of quantum mechanics.

In double slit experiments, a beam of particles (or photons) is aimed at a barrier containing two slits side by side. The beam is comprised entirely of particles of a single type. Each slit is large enough to allow through the particles being studied. Particles that pass through the slits hit a screen that emits light where the particles hit and are thus observed. Individual particles generate dots on the screen. Beams of particles create patterns on the screen. The pattern that appears on the screen depends on the distance between the slits in the barrier. Classically, particles that hit the screen behind the barrier should pass through one slit or the other. In the actual experiment, when the distance between the slits is sufficiently large, the pattern created by particles passing through the slits is the same as the classical prediction. The phenomenon of interest occurs when the distance between the slits is within the range of quantum uncertainty. In such cases the pattern on the screen changes as if, somehow, even though each particle must go through one slit, its path is affected by the presence of the other slit.

The hypervortex model provides a nice simple physical explanation for this otherwise strange behavior of double slit experiments. In reality, it is not particles passing through slits; rather it is vortices passing through hyper-slits. Each vortex, due to its variable path (especially due to the surplus-length-enabled waves along its length) sees both of the slits and sends some energy through each slit. As the vortices pass through the slits, their long length is broken into short segments, some segments passing through each slit. However, a set of short segments is not a preferred structure in the hyperfluid. After the short segments pass through the slits, their preference is to reform into long, stable vortices, as was discussed in an earlier chapter. That activity, the rejoining of short vortex segments into long vortices creates an attraction among the short vortex segments. It is that attraction that changes the pattern on the screen.

In contrast to the hyperfluids rather simple physical explanation of these experiments, the explanations provided by other models caused Einstein's disbelief of the theory. Hence, for example his comment "God does not play dice with the world".

Entanglement: Entanglement is the term given to the phenomenon underlying occurrence of Einstein's "Spooky Action". This section discusses in some detail, the occurrence of the underlying entanglement that enables it. Again, "Spooky Action" is when some action taken on one particle affects another particle some distance away instantaneously, or nearly so. This does not happen to all particles, only to particles that are mutually entangled. Entangled is the standard term for the coupling between two particles in such a case. The physical explanation of the entanglement is very different in the hypervortex model than otherwise. In fact, in most models there is no physical explanation of the phenomena, only the necessity of its existence to provide consistency with experiment.

First we will discuss what entanglement is not. Despite the term "entanglement", it is not simply a result of several vortices becoming entangled. Recalling the analogy of a kite on a string to a vortex in the hyperfluid, consider two kites on strings (two vortices) when the strings cross. With kites on strings, two or more kites can become entangled. If they do, they affect one another's motion. Also, vibrations of the various kites and strings will couple and the vibrations will travel between kites across the strings at speeds much higher than the speed of the wind. In the analogy of kites to particles and their strings to the corresponding vortices, we would only see the kites and the "spooky" coupling of the actions between them. However, vortices are not kite strings. When kite strings tangle, each string says intact. In contrast, vortices are more pliable and adaptive. As the double slit experiment shows, if something creates an obstacle to the motion of a vortex it splits up in pieces to go around the obstacle and then reconstitute itself. If vortices cross and tangle like kite strings, they quickly reorganize to untangle so that each particle can again be untangled and behave as untangled.

Entanglement of two particles that allows "Spooky Action" is a special case. It occurs when a vortex is created that crosses through our observable universe twice. Such creation occurs in events that generate a particle/anti-particle pair (as discussed at the end of Chapter 2). The particle and anti-particle are two observations of one vortex. In the analogy to kites, it is like a string with a kite at each end, and the kite flyer holding the string at a point halfway between the two kites. If an action to one kite creates a vibration, that vibration travels down the string, past the kite flyer and up the other string. With actual kite string and a kite flyer the vibration might dampen along the path, but with actual vortices the level of such damping is much less or nil. The distance that the vibration travels along the string to go between the kites is much longer than the distance between the kites. However the speed of sound along the string (which is the speed of the vibration in the string) is so much higher than the speed of the wind that the fastest path for information flow between the kites is still along the string. Similarly with vortices created in events that create particle/anti-particle pairs.

There is a major difference between kites on strings and particle/anti-particle pairs. To examine it, consider a different analogy, one that shows both the differences and how entanglement can occur with photons. Consider a boat in motion creating a wake. Also, imagine viewing this boat and its wake from above, but through a slit that allows seeing only a long narrow slice of the top of the ocean. The boat passes by in a direction perpendicular to the slit. Thus, we view the boat only briefly. The boat is analogous to the moment when we observe the creation of a particle/anti-particle pair or a similar photon pair. The boat in motion, as usual, creates a continuous wake. However, viewed through the slit we see only two separate waves emanating from the point where the boat passed by and travelling in opposite directions. Observation of the two waves in the water is analogous to our observation to two photons in the hyperfluid emanating from a creation event. Despite the fact that we have two separate waves they really are part of the same wake.

In the hyperfluid, the boat represents an event that causes emission of the entangled pair of particles or photons. The slit represents our ability to see only one observable universe of the larger Hyperverse. The two waves that we see emanating from the boat are like two photons emanating from the event. The two photons look separate but are coupled by a hyper-dimensional structure being created as the generating event (the boat) propagates across many other observable universes in the Hyperverse. If something acts on one of the waves, depending on the type of action and depending on the properties of the hyperfluid and the structure of the wake, information can possibly travel as a vibration in both directions along the wake. The vibration traveling in one direction will travel back to the boat and then down the other side of the wake affecting the behavior of the photon that we observe on the other half of the wake. For certain types of vibrations (likely longitudinal vibrations) the speed at which the vibration travels will be much faster than the speed of the wake and thus we will observe "Spooky Action". It is not clear whether anyone has studied when and whether wakes in water have vibration modes that would create a coupling between two sides of a wake, but that is not the point. Rather, the point is that in the hyperfluid, the wake-like structures created by such events, and the vortex structures created by other similar events, do have vibration modes that couple the two separate observed particles (or photons), with a high speed connection that we observe as Einstein's "Spooky Action".

6. Origin of the Thesis

The "Aha" Moment: The origin of this thesis began with a real "aha" moment. However, that "aha" moment was built on much prior knowledge including both knowledge of prior models and knowledge of various opinions of those prior models. That knowledge guided this effort away from directions that had already failed and into directions that had been considered interesting but which in some aspect had been overlooked.

A universe filled with ideal zero-viscosity fluid (and variations) was a mainstream viewpoint of scientists for a long time. A fluid was postulated by Renee Descartes in 1644 as the mechanism underlying gravity. Isaac Newton attempted his own variation of a fluid model of gravity some decades later. Leonhard Euler (of mathematics fame) attempted such a model. James Maxwell (author of Maxwell's Equations that unify electricity, magnetism and optics) postulated a fluid as the medium in which light travels and as the underlying phenomenon unifying electricity, magnetism and optics. Many additional scientists, famous within the physics community, postulated various fluid structures and behaviors to explain the root cause of gravity or electric charge. The Higgs boson, postulated in 1964 and for which the Nobel Prize in physics was recently awarded, is conceived as an observable phenomenon rooted in an invisible field that permeates the universe, and which enables particles to have mass. (Is that field the hyperfluid?)

Throughout the long history of prior fluid models, it was generally not postulated that the particles were vortices in the same fluid. At the time no one had seen a view of a hurricane as a vortex – there were no pictures from satellites in orbit. No one had seen a video of a tornado – there were no videos. No one had put Helium3 into a super-fluid state with zero viscosity and containing vortices. Also, much less was known about elementary particles; there was no basis to map various particle types to various vortex types. Strong and weak forces had not been measured. There was no concept of wave-particle duality. No one had seen a particle act like a wave; such experiments occurred later. Lord Kelvin (of temperature-scale fame) was a notable exception in attempting, late in the 19th century a model of atoms as smoke rings in the fluid, a type of vortex that had been seen in air in that era. Still, lacking detailed knowledge of particles or vortices, a correct construction was beyond his possibility at the time.

The absence of a view of particles as vortices in those fluid models led to a view of particles as a separate material in the fluid, which gave rise to the Michelson-Morley experiment discussed in Chapter 1. The results of that experiment led to Einstein's efforts, which led the physics community in a direction away from fluid models, but without having explored the possibility of particles as vortices.

Many believe that Einstein proved that the fluid model was wrong, but such belief is historically and technically wrong. Einstein's three postulates, in Chapter 1, are all properties of fluid. Consider specifically his postulate that the speed of light is independent of the velocity of the emitter. The same is true of sound in air, or in any other medium. The speed of waves in any material is independent of the velocity of the object initiating the wave. Consider also the principle of relativity. Galileo developed the principle and used sailing on the ocean as his example – a person inside the ship cannot tell the speed of the ship, only the choppiness of the sea state. There is nothing about Einstein's postulates that is at odds with properties of fluids.

Einstein was not telling people that the fluid concept was wrong; he was using the results of the Michelson-Morley experiment to determine the correct properties of the fluid. Einstein was telling people, in part, that instead of guessing properties for the fluid and wondering why the data disagreed with their results, rather people should use the observed data to determine the properties of the fluid. He then presented results of doing so. The hypervortex model follows that perspective of Einstein's work..

The above history told me where to look for a new model, but it was Dirac, who truly told me to seek a new model. Dirac, the person who first combined special relativity and quantum mechanics and predicted the existence of anti-particles, and likely the last recognized surviving founder of quantum mechanics chided, late in life, that modern physics theories had not provided the level of understanding that he had sought and hoped for. Some years after Heisenberg, the last other surviving co-founder of quantum mechanics, had passed away, Dirac, at a small seminar, face-to-face, presented to us, to me, the notion that we, collectively, do not understand even the electron. He did not have an alternate theory to propose. He did not discuss his accomplishments or those of his contemporaries, except to emphasize the need to replace their work with something better. He had taken the time and the trouble to speak, and he chose as his topic to tell us, to tell me, in no uncertain terms that we need a different approach, an approach that provides deep physical understanding as well as mathematics. This work is my attempt to respond to his request.

The "aha" moment comprised a realization of the existence of a filter that we as observers place between us and the real world. In particular, I noted how strongly we rely on detecting constants. No matter that something may be rapidly varying such that the observed constant is really an average over time, we observe a constant and then treat the universe as if that aspect really is constant. Realizing that this filter existed, I imagined what reality might look like with this filter removed. The reality that subsequently appeared to me was hypervortices undergoing complex hyper-dimensional motion that we observe as particles in proper relativistic motion. Exploring and describing that reality with mathematics has occurred incrementally.

Increment 1 – Special Relativity in a Hyperfluid: The first step towards development of a hyperfluid model was to determine whether the addition of a dimension could allow construction of rules of motion for hyper-particles that is consistent with special relativity, but without the need to mix space and time. This achievement occurred somewhat accidentally. While there was a belief that the fluid model concept should be revisited, there was no particular idea that hyperspace would achieve this goal until this increment of the model development was already achieved. That is, it was the achievement of this increment that really launched this effort to create the hypervortex model.

Consider the idea of hyper-particles as the extension of particles into hyperspace without regard to the underlying cause for their existence. In that sense, vortices in hyperfluid are a particular way for hyper-particles to exist. The work in Increment I of the effort involved several questions. First, could hyper-particles, however comprised, move in a way that causes us to observe relativistic motion? Also, could they move in a way that would cause us to be able to not observe the additional spatial dimension? Such work began, on this effort, in about 1980. Others had constructed models of particles in extra dimensions and found that under their assumptions, they could not answer these questions in the affirmative. To read about those other efforts, one might search "world-lines" on-line in articles, books, etc., that discuss relativity.

For this work, the effort reversed the approach that was used to develop the concept of world lines. Instead of assuming rules of motion for hyper-particles and checking whether those led to the desired answers to the questions, the effort used Einstein's approach. It started with the observed properties of the rules of motion, and derived hyper-particle rules of motion that fit with the observations. The results showed that hyper-particles could move in ways that cause our observation of relativity and our inability to observe the extra dimension. It was not mathematically hard. The addition of an extra spatial dimension provided an extra variable, an extra degree of freedom, which was adjustable to make the rules of motion match the data. Einstein derived the properties of relativity using dimensions (x, y, z, t) and in order to do so he changed the prior standard meaning of those separate space and time dimensions into one in which the concepts of space and time mixed. The Increment I hyperfluid work used dimensions (w, x, y, z, t) which achieved the same overall physics without a need to couple space and time.

The effort required for this achievement was not hard or long once the "aha" moment had occurred. The results were presented at an American Physical Society conference to a large audience (standing room only), followed by enthusiastic discussion at an impromptu lunch arranged afterwards by a member of the audience. However, journal editors of the selected physics journals were less enthusiastic.

The established community was not ready to allow the standard models to be threatened, just because the possibility that a fluid model could work had been restored.

Increment 2 – Gravity and Curvature in the Hyperfluid:
Once Einstein had published his special relativity, he worked to generalize it. That meant adding the effects of gravity to his model. In doing so, his coordinate system not only mixed space and time, it also curved space and time. The mathematics became very complicated. Increment 2 of the development of the hypervortex model followed Einstein's goals, that is, it generalized the results of Increment 1 to include gravitational forces but without curving space and time. Increment 2 of this effort, thus, was to determine the specific properties of a hyperfluid that would reproduce the key confirmed observable predictions of Einstein's general relativity. This began with mapping together Einstein's postulates and Maxwell's fluid properties. Such analysis found that Einstein's postulates amount to the following refinements to the fluid: (1) a frictionless mechanism exists in the fluid that resists velocity gradients without viscosity; (2) particles are comprised of fluid; (3) an extra spatial dimension (or dimensions) exists that the fluid fills, and (4) the fluid is a compressible hyperfluid such that its changing density causes effects such as gravity.

Increment 2 created a level of equivalence between the fluid approach and the geometric curved-space-time approach used by Einstein to develop General Relativity. In Einstein's mathematics light travels a straight path in a curved space; the curved space is caused by mass, and the curvature of space causes gravity. In the hyperfluid, light travels a curved path in a straight (flat) space; the curved path of light is caused by changes in the density of hyperfluid around mass. Then, depending on one's preference, one can say either that the curved path of light causes gravity or that it is a manifestation of the gravity caused by the variable density of the hyperfluid. The idea of equivalence between light traveling along straight paths in a curved space or along curved paths in a flat space is not new. It was attempted by others when Einstein first published his works. What was new in Increment 2 of the hypervortex model development was the achievement of the equivalence, which required a hyperfluid rather than a 3D fluid.

Equivalence of Einstein's mathematics and the hyperfluid mathematics means that both give similar answers for those things that both approaches can discuss. The mapping shows, contrary to historical pretense, overlap, not divergence, between the fluid and the geometric views of physics. A final step to achieve correct results in the hypervortex model was to realize that because measurement instruments are made of fluid (if particles are made of fluid then everything made of particles is made of fluid) hence all observations must be adjusted for the effect of the fluid on the measurement instruments.

The ultimate value of models is their ability to predict science beyond that which has been measured. That is, while it is nice to have models explain results after the results have been measured, the better indicator of a model's value is whether it can predict the correct answer before it is measured. The model that resulted from Increment 2 of the hyperfluid development was used to predict and explain a divergence in the results for two different ways of measuring the age of the universe. One method to estimate of the age of the universe is from measuring the rate of expansion of the universe. If we know its size and its rate of growth, we can compute its age as the length of time required to expand to the current size. The other method to estimate the age of the universe is from the rate of development and evolution of galaxies; which requires a model of their development. The fluid model was used to predict that these two methods would provide very different answers, and subsequent publications did indeed confirm that the two methods did provide very different answers. The difference was caused by the fact that the rates of clocks have varied over time since the beginning of the universe expansion, which was not included in the models of the day. Particle models still do not include the variable rates of clocks in their computations of the age of the universe; however particle models have infinite adjustable parameters which were used then to make the two estimates match well enough to avoid a crisis for the particle models. Particle models of the universe had no real ability to predict the age of the universe, but with infinite parameters they could be adjusted to explain it. However, even with all their adjustable parameters they could not explain other effects however, such as a universe whose expansion rate is accelerating, or how galaxies actually form since they contain insufficient mass to coalesce into spinning spiraling discs.

Eventually, the results of the first and second increments of the hypervortex model were published on-line when the world-wide-web opened up such possibilities.

Increment 3 – Unification in the Hyperfluid: While Increment 2 of the hypervortex model development provided a mathematical framework to unify physics many details had not then been incorporated into the framework. Increments 1 and 2 of the hypervortex model development had repeated Einstein's work in a fluid model, but they had not gone much beyond. They had not unified gravity with other physics. Increment 3 has achieved that goal. The mathematical framework has been sufficient for the effort. That is, there is still no need to curve space or to mix time and space in the hypervortex model. The hypervortex model has been successful. This work provides select essential aspects of the model. Additional mathematical proofs that further validate the model are being prepared for separate publication.

The main mathematical differences between the hypervortex model and Maxwell's 19th century Equations for electromagnetism is shown later to be: (1) the addition of one equation to his set of equations, and (2) the ability to derive all of the equations from one master equation. Other minor refinements to his set include: (1) reference to an additional dimension in his equations and, (2) an auxiliary equation that defines in detail the relation between electric charge and its cause in the hyperfluid. Maxwell's model, it turns out, already had most of the necessary ingredients. Still, the additional equation directly incorporates gravity, and quantum mechanics. Also, it integrates nuclear forces via its enablement of the derivation of the internal structure of elementary particles (vortices in the hyperfluid). In doing so, it exposes the reasons that particles have the properties that they have. It shows that Maxwell and Einstein were both almost right about the nature of light. The new equation, combined with Maxwell's Equations, shows how light is both a wave and a particle.

The geometric mathematics of General Relativity (and other variants) is not reused in this model. It probably could be reused for certain physics. One value of a fluid model's mathematics over the geometric mathematics used in particle models is that the fluid model has larger scope, deeper insight, easier interpretation, and easier use. For example, General Relativity uses very complex math to avoid referencing any point in space outside our observed universe. As a consequence, it lacks the scope needed for expanded discussions of the universe; it is very difficult in such mathematics, for example, to discuss the motion of our observable universe within a hyperverse. The presence of a fluid provides a unifying physical mechanism for why all things happen. In comparison, geometric models do not even explain why gravity curves space.

To be clear, Einstein's postulates are central to the success of the hypervortex model. Without them, this fluid model likely would not exist today. His postulates guide the update of Maxwell's fluid model to create the hypervortex model. Also, to be clear, Einstein initially rejected Minkowski's geometric interpretation of relativity. Einstein's postulates for special relativity were presented at a time when a fluid model was dominant and as refinements to that view.

Of the various types of contributions – Maxwell's, Schrödinger's, Dirac's, and Einstein's – this work seems most like Maxwell's. Maxwell selected among the work of Gauss, Coulomb, Helmholtz, Fermat and others. He then added a known but overlooked component and output an integrated result using well-known, reliable mathematical formalism. Schrödinger, in contrast, made a leap of mathematics to adapt the Hamiltonian approach to the mathematics of the physical world (discussed in Chapter 8) and thereby created a solid equation and theory for the study of quantum mechanics. The deep significance of his work becomes clearer in the mathematics of the hypervortex model. Dirac achieved a leap in the unification of relativity and quantum mechanics, and made breakthrough predictions. Einstein contributed multiple far reaching physics concepts that have changed the world, but whose mathematics remain disjoint even today.

Sorting through the available pieces of theories and models, as Maxwell must have, has been pedantic work. The effort to build the hypervortex model began at a time when disjoint pieces abounded, and new observations were challenging the standard models, and the standard models could not answer the expanding set of questions that were and are important to people. The selection of parts to keep, parts to transform, and parts to discard has been an iterative process, guided by Dirac's request for a model that provided deeper physical insight and a less mathematically intense foundation.

The legacy model component whose decision to keep has been easiest is Lagrange mathematics. Lagrange mathematics is named for its author. Formulation of a unified theory using his mathematics has been a goal of the physics community for generations. Lagrange mathematics provides the simplest and most widely used form for generating new physics models, and it is widely used in engineering. Conceptually, it simplifies analysis of systems to a minimization problem. The formalism was matured by Euler (1707 to 1793) and then Lagrange (1736 to 1813) to achieve a precise and general mathematics applicable to all particles in all environments. The principle was promoted to its current status by Hamilton (1805 to 1865). Thus, it was mature before Maxwell used a fluid model in the 19th century to unify electricity, magnetism and optics. This stalwart workhorse of physics has not required fundamental update since then.

The mathematics of Lagrange allows representation of the kinetic and potential energy of a system in ANY generalized coordinates. Indeed, the mathematics includes a formal proof needed here that any coordinate system is valid in which the kinetic and potential energy of the system can be written. Like a magical black box, it takes that generalized representation as input, and it outputs equations of motion for the system. Lagrange's mathematics and the precise definition of Least Action are central to the results of the hypervortex model as shown in Chapter 8. Lagrange mathematics includes a feature that allows for clocks whose rates vary. That is, it includes a concept of time and a separate concept for the rate at which things age. Moreover, it provides mathematics that computes the relation between the two as it depends on the local physics of the universe. These features of the Lagrange mathematics makes it very appropriate for development of the hypervortex model. For contrast, note that most models today assume a constant one-to-one relation between time and aging. Thus they actually are in conflict with features of the Lagrange mathematics. Still those models use the Lagrange mathematics, the historical result of which has been the convolution of the coordinate system to force fit their assumption to the Lagrange mathematics.

The other prior mathematics that turns out to be completely reusable in the fluid model is Schrödinger's equation for quantum mechanics. In the domain for which it was designed – small non-relativistic perturbations – it seems to need no changes to work with the hypervortex model. The hypervortex model helps show why Schrödinger's Equation works, and what equations to use instead when it is not valid.

The legacy model component whose discard was easiest was Minkowski's curved space-time coordinates. The decision has been affirmed by many factors. First, Einstein's whole concept of relativity fundamentally indicates that the physics should work in any choice of rest frame. Therefore, it is valid for the hypervortex model to use a rest frame in which observable universes are in motion in a larger stationary hyperverse and to thereby better study their existence and motions. Also, therefore, there is no requirement to use Minkowski's complex curved space-time to avoid referencing locations in a larger hyperverse. Second, Lagrange mathematics proves that any coordinate system that can represent the physics is valid so, simple flat coordinates that separate time and space seem best. Third, not only did Einstein not create the geometric view; he initially rejected it. Fourth, Einstein could never truly finish his theories using it, and since he could not, then likely neither could I, in that mathematics. Fifth, only a miniscule number of people are capable of applying Minkowski's mathematics to physics. Having, myself, aced the necessary graduate level mathematics courses, I still found the mathematics distracting from the physics. Thus, since Lagrange had proved that alternatives could be used, such alternatives seemed best for modeling the hyperfluid.

Observations on the Transformation Process: The decision to discard Minkowski's mathematics was reaffirmed by something that I learned much later. Einstein made many mathematical mistakes in his published works. His errors fill at least one published book dedicated to the subject. These mistakes were not individual occasional errors. On some topics he made more than a half-dozen related errors in different articles on the same or similar topics. This is not mentioned here to denigrate Einstein. Rather, Einstein is a standard bearer for the definition of genius. His mistakes affirm that such mathematics is far too complicated for use in physics.

Einstein is quoted as saying "Everything should be made as simple as possible, but no simpler". The quote is likely a paraphrasing of his actual statement, "It can scarcely be denied that the supreme goal of all theory is to make the irreducible basic elements as simple and as few as possible without having to surrender the adequate representation of a single datum of experiments". Using either quote, mathematics is an element of the theory and, if Einstein made repeated mistakes using the mathematics of curved space time, a simpler mathematics is required. In fact, using Einstein's quote, dubbed by others as "Einstein's razor", everyone should prefer the simpler mathematics of the hyperfluid over the much more complex mathematics of other models.

Historical background, as provided in university physics curricula, was important to the effort. Without that background, I might not have known of Maxwell's fluid model. Also, I might not have known of the litany of experiments whose results would have to be reproduced by a new theory. Further, I might not have known of the long list of already tried paths to unification and their outcomes. Without that background, the hypervortex model would not exist.

To augment the historical background provided by the university, the effort included a scan of the history of physics, mathematics, astronomy, and more back to the ancient Egyptians, Sumerians, Phoenicians, etc. That search revealed examples useful towards deciding whether Maxwell's concept of fluid, having been rejected, should be resurrected, and if so, how. The history survey revealed things that I now think most of us should know from our schooling, but do not. For example, Darwin in espousing a theory of evolution was espousing a concept already known and well discussed and believed in the time of Aristotle. Also, Galileo's espousal that planets circled the sun was fact in the time of the Pharaohs, as was the fact that the Earth was round. And, atomic theory as it evolved from Rutherford's experiments was in part a resurrection and refinement of an ancient Greek theory, which though popular in that time was rejected by Aristotle. If such important elements of modern science could have been known and discarded, and then resurrected millennia later with such benefit as we have obtained from them, then the fluid model of the universe, discarded for a mere 100 years, could also. That helped me to decide to give the fluid model a chance.

Analysis of the most famous and most historically referenced bit of conflict in the physics community strongly contributed to the effort. The particular historical conflict, of course, was the argument over whether the Sun revolves around the Earth or vice versa. If Einstein is right that it is all relative, why does it matter? Why argue over which is the center, or which is at rest, if both views are actually correct? The hyperfluid provides clarity on why it does matter, on what is relative and what is not relative. Newton noted that the universe might be relative or absolute. He also indicated that he lacked data to determine the truth. Einstein did not prove that everything was relative; he made an assumption. The hypervortex model shows that things are relative as long as one only asks questions about the local region of space – the region nearby where the fluid is at rest relative to the piece of the universe about which a question is being asked. However, when a question regards a larger region of space across which the fluid's density and motion vary, then knowledge of the fluid's motion is required to get the actual correct answers. In such cases, selection of the correct object as the center of the discussion does matter. The Earth can be used as the center of the discussion for questions regarding events for which the fluid is at rest relative to the

Earth. The Sun is the correct center of the discussion for questions regarding events for which the fluids motions are determined by it. And, as we expand the discussion, the galaxy, etc. each becomes the better center for discussions in which the fluid's motions are governed by the larger body. This clarity is important to the construction of the model.

Early Responses: When this model was in early development and key accomplishments showed the basic validity of the fluid model, I presented a draft to a subset of the professors at my alma mater, the university from which I received my doctorate degree. Some of them had particular comments that have affected the effort.

One of my physics professors had a favorite refrain relevant to the effort, "Today's heresy is tomorrow's orthodoxy." Besides telling me that what happened to Galileo, and the whole argument of whether the Earth revolves around the Sun, could happen today; it implicitly told me to count the heresies contained in the hypervortex model. Originally there were three: (1) Speed of light is not constant, only observed as such, (2) There is an unseen dimension to the universe, and (3) Maxwell's fluid is correct in its essence and all the physics arguments made to bury it have been wrong. The theory has not changed in these regards. Fortunately, fewer of these statements are considered heresy today.

Another professor's response was a question, "Have you read the book, Flatland?" He was referring to Flatland: A Romance of Many Dimensions. It was written before general relativity was conceived and gained its importance afterwards. It is an illustrative tale about scientific heresy presented via the difficulty of convincing a society of the existence of dimensions beyond those observable to that society. While its intent may have been using dimensionality as an example, its popularity in the physics community remains specifically about conceiving and believing in dimensions beyond those observed. The hero is a 2D being who lives in Flatland, a place where suggesting the existence of a 3D universe is heresy. The hero gets to enter the 3D world via a visit from a 3D being, but, after suggesting the possibility of a 4D world, is ejected back to his 2D world for heresy. Despite the popularity of the book, actual suggestion of a real 4D universe was still heresy at the time of my draft. Today, suggesting extra dimensions to the physics community is less heretical. Still, most theories with extra dimensions have little balled up dimensions that are constructed so as to be too small to observe. The existence of a dimension large enough to observe but still unobservable to man is still less acceptable among many.

The professor who was head of the university physics department at that time responded to the theory with, "People will listen when they are ready to listen." Periodically, I have asked myself whether enough people are ready to listen. At times I have found that enough are willing that I should keep up the effort, but from a position in which my success does not require that people listen. Thus, I avoided the academic career path and the threat of "publish or perish". Instead I have spent more than thirty years as a scientist for various endeavors, which in fact has provided opportunity to explore the breadth of physics involved in a theory of everything. That same professor requested more clarity regarding the assumptions required by the model. He was thinking of "Einstein's Razor" and wanted a presentation that would help people compare the hyperfluid to standard models in that regard. The next chapter is a direct response to his advice.

7. Fewer Postulates

The fewer the postulates a unification model requires, the better it is, in general. Whether one uses Einstein's Razor, Ockham's Razor, or some other phrasing of the basic goal of the model development, the goal is simplicity that achieves completeness. Using Einstein's phrasing, postulates are the irreducible basic elements to which he refers in his original, unabbreviated rendition of "Einstein's Razor." Per his statement these postulates should be as few and as simple as possible, while still explaining the universe. This chapter provides a postulate-by-postulate comparison of the standard particle models and the hypervortex model. It shows the hyperfluid is, by Einstein's criteria, much preferable over those standard particle models. The hypervortex model has fewer postulates, and the definitions are more specific and hence clearer and simpler.

In practice, counting the postulates can become a matter of semantics and a blunt instrument for judging models. For example, is assuming that the universe has three spatial dimensions more or fewer postulates than assuming that the universe has four spatial dimensions? Is each dimension a postulate? Here a postulate is anything input into the model, which, if wrong, will make the model wrong. A good model uses mathematics that allows the number of dimensions to be determined by a comparison of the model to the real world. Such a model makes no assumption about the number of dimensions. Thus, the model cannot be made wrong by such an assumption. Rather, with such an approach, the model tells us how many dimensions are needed to describe the real world. Hopefully, by using such an approach to the identification of postulates, this section provides clarity.

Translating Postulates: Table 1 enumerates key physics postulates from the standard models whose status changes in the hypervortex model. Column 1 lists the legacy postulates, and provides, by itself, a nice summary of the state of physics theory prior to the hypervortex model. It provides a set of things that a new model must explain, postulate, derive or replace. Column 2 of the table indicates the status of the each standard postulate in the hypervortex model. Column 3 adds some detail to the status provided in Column 2.

Table 1. Mapping Postulates from Standard Models to the Hypervortex Model		
Standard Postulate	**Status in the Hypervortex Model**	**Commentary**
Particles are the elemental building blocks of the universe	Replaced	Hyperfluid is the elemental building block of the universe.
The structure of the universe is built bottom-up from collections of particles.	Replaced	The structure of the universe is largely a top-down hierarchy of stable structures formed of hyperfluid and which then interact.
Newton's Principle of Inertia	Translated	Newton's principle of inertia implies an ideal hyperfluid with no viscosity and no loss to friction, and a non-viscous mechanism that limits velocity gradients in the hyperfluid.
Newton's Equal and Opposite Forces	Translated	Translated here to the postulate that hyperfluid is continuous, from which Newton's force laws are derived here.
Principle of Relativity	Translated	Wherever one is, one is at rest relative to the local fluid.
Wave Particle Duality	Translated	Implies particles and light are both composed of hyperfluid.
Speed of light is independent of the velocity of the emitter	Translated	The hyperfluid is compressible. Had Einstein said the speed of light is a constant, that would have implied a non-compressible fluid.
Maxwell's Concept of Continuity	Translated	Implies continuous hyperfluid; and conservation of hyperfluid, i.e., that it is neither created nor destroyed within the context of our existence.
Four Forces mediated by exchange of particles.	Translated	Forces are the interactions among vortices.

Table 1. Mapping Postulates from Standard Models to the Hypervortex Model		
Standard Postulate	**Status in the Hypervortex Model**	**Commentary**
Physical Constants, h, e, g, c, ...	Translated, partially derivable.	The potential energy terms in the hyperfluid Lagrangian, when correct, allow derivation of the measured values for these physical constants.
Inertial mass equals gravitational mass	Derived here	For vortices that have mass, the size of the vortex (which determines gravity) is shown to be proportional to the energy of the vortex (which determines it's mass). Then, by convenient selection of the gravitational constant, inertial mass and gravitational mass are equal.
Gravity is additive	Derived here	Whereas other models need to assume that the gravitational force created by a collection of particles is additive, this property of gravity is derived in the hypervortex model.
Heisenberg Uncertainty	Translated, eventually derivable	Vortices have length in excess of that needed to follow the shortest path through space over time. The excess length allows waves to form in the vortices that allow the path to deviate from the classically computed path. Heisenberg's uncertainty equation translates as the amount of excess vortex length.
Quantum Entanglement	Translated, eventually derivable	Entangled particles are two independent observations of the same vortex. "Spooky Action" results from waves propagating along the vortex between the two observed particles.
Space-time mixing	Deleted	Rather than mix space and time, a spatial dimension is added to the fluid to make hyperfluid, while keeping global time separate.

Table 1. Mapping Postulates from Standard Models to the Hypervortex Model		
Standard Postulate	**Status in the Hypervortex Model**	**Commentary**
Constancy of the relationship between clocks and time.	Deleted	Clocks measure Action. The relation between action and time varies. That relation is usable to convert a clock's output to a time.
Dimensionality of the universe.	Deleted	The hypervortex model stops assuming that we observe all the dimensions of significant extent that exist; rather in the hypervortex model we derive the reason that we observe three spatial dimensions.
Curvature of space	Translated and Derived	In the hyperfluid the coordinate system is flat; observable universes are curved moving sub-spaces in the hyperfluid as a consequence of the hyperfluid's motion as affected by its compressibility.
CPT (charge-parity-time) Invariance	Derived	Automatic property of ideal, continuous, hyperfluid in which charge and parity are properties of vortices.
Parallel Universes	Derived	Parallel universes derive from the kinematics of the hyperfluid.
Matter formation after the big bang	Translated	In the hypervortex model, matter formed from turbulence in the expanding hyperfluid.
Dark matter to explain how gravity can hold galaxies together.	Deleted	In the hypervortex model, galaxies are ultra-large scale vortices that hold themselves together just as smaller particles (e.g. electrons and protons) do. The objects that we do see inside galaxies are smaller vortices trapped by the gravitational field of the galactic scale vortex.

Table 1. Mapping Postulates from Standard Models to the Hypervortex Model		
Standard Postulate	**Status in the Hypervortex Model**	**Commentary**
Dark energy to explain observing an accelerating expansion of the universe	Deleted	In the hyperfluid observation of an accelerating universe can result from any of three attributes of the hyperfluid with no need for a special postulate.
Constant size of objects as the universe expands	Deleted	In the hypervortex model, we stop assuming a particular scaling of objects as the universe expands; rather we use the properties of the fluid to determine the actual scaling of objects, which turns out to be that object size is not constant, only observed as such.

All postulates needed for the hyperfluid are obtained as translations of an existing standard postulate. In column 2 of the table, postulates are tagged mostly as "replaced", "translated", "derived", or "deleted". Those marked "replaced" are foundational to the transition from a particle view of the universe to a fluid view of the universe. Those marked "translated" are used to construct the hypervortex model such that it has the right properties to explain observed phenomena. Postulates marked "derived" and those marked "deleted" are not postulates in the hypervortex model and reduce the number of postulates in the fluid model. For those marked "derived", the scope that is postulated in the standard models is within scope and explained by the hypervortex model, however in the hypervortex model the information is derived from other postulates, not postulated. Postulates marked "deleted" are fundamentally wrong explanations of the physics or at least misleading in the hypervortex model and need no replacement. The few marked "derived here" have derivations performed for this effort and included in Chapter 8. Postulates marked "translated, eventually derivable" use a postulate now, yet the model seems to provide scope that will allow future derivation of the postulate. Such future derivation, if achieved, will further reduce the number of postulates needed for the hypervortex model.

The construction of the hypervortex model has deleted or derived many standard postulates. Consequently, the hypervortex model has many fewer postulates than standard models. It also has larger scope. Thus, defining value of a model as the scope per postulate, the hypervortex model has much more value than the standard particle models that it replaces. "Value" is an important metric for a model. This metric should be used more. It is the type of metric that is used in the commercial industry to determine the relative value of many things. It is a measure of return on investment, and the return on investment for the hypervortex model exceeds that of standard models.

Simpler Postulates: Per Einstein's Razor, the relative simplicity of the postulates is a factor to consider in comparing models. Just as the counting of postulates includes potential for semantic confusion, so too is the judgment of simplicity. A postulate may sound simple when read, but if its meaning is ambiguous then it is far from simple. Also, it may be simple from a qualitative view, but if it lacks the precision required for incorporation into the mathematics, then it is far from simple. The postulates of wave-particle duality and Heisenberg Uncertainty exemplify the difficulty. For both postulates, the hypervortex model provides the simpler, quantified postulate.

Wave-particle duality in the standard models provides qualitative guidance on the similarity of particle and wave behaviors. However, in the standard models, there is no specific equation for the concept; there is no specific place where the postulate integrates with the other aspects of the model. In contrast, in the hypervortex model, the same postulate translates simply and clearly to state that both waves and particles are composed of hyperfluid. The hyperfluid's version is simpler and clearer to use than the standard version of the postulate.

Heisenberg Uncertainty exemplifies a postulate with quite a different difficulty. This postulate is a specific equation in the standard models.

Its limitation is, however, ambiguity in the meaning of the terms in the equation. Its equation provides a relation between the uncertainty in particle location and particle momentum. Interpretations of it have varied over time, and across physics communities. Its interpretation is a subject of substantial discussion in physics classes. Some interpret it as a fundamental property of the particle, others as a limitation on the ability to simultaneously measure both momentum and location of the particle. A presentation of all such interpretations and arguments and the supporting experiments is a book until itself, or could be. In contrast, the equivalent postulate in the hypervortex model is very specific in saying that the equation indicates vortices have more length than required to connect two points. This surplus length allows the vortex to deviate from the classical path. The terms in the equation then describe and constrain the deviations that the vortex can take relative to the shortest path. That definition leads in Chapter 8 towards a future ability to derive the source of these detailed vortex behaviors, which eventually is expected to replace any need for a postulate on this topic.

A Stable Set of Postulates: A third factor important to judging a model's value is exposed by examining the history of the addition of postulates to the standard model. The standard models have been patching-on *ad hoc* postulates in increasing numbers to accommodate data that disagrees with the model. Moreover, the addition of each postulate adds no other value to the model other than to repair its disagreement with that data. Also, the new postulates have not been increasing the scope of the model. Thus, the value of the standard model in terms of physics explained per postulate has been declining.

In contrast to the current set of standard models, the value of the fluid class of models increases as it evolves here from the Aether model that Maxwell used to derive electromagnetic theory to the hypervortex model presented here. Maxwell's equations are a product of the fluid class of models. His equations explain and unify optics, electricity and magnetism. That implies very good value for those equations. Now the hypervortex model, via addition of one equation and removal of a constraint, adds the integration with gravity, strong and weak forces, and quantum mechanics. Thus the evolution from Aether to hyperfluid at least doubles the scope of the fluid class of models. Such a large return on investment has been the essential reason for investing the time to develop the hypervortex model.

While the standard models have required frequent patches, over the same period the hypervortex model has required no patches. New data that has challenged standard models has posed no new challenge to the hypervortex model. In Table 1, postulates like Dark Matter, and Dark Energy are not needed in the hypervortex model. All the data that drove the addition of these unobserved unknown object types to the models required no changes to the hypervortex model.

A Foundational Change of Perspective: The first two postulates discussed in Table 1 provide a fundamental shift in man's perspective of the universe. For much of the past 100 years physicists have worked from a perspective in which the universe is fundamentally comprised of particles and that everything we observe is essentially built bottom-up from small particles into large objects. For the understanding of small-scale physics this perspective has been very valuable, but beyond a certain scale it has repeatedly failed.

For physics that occurs on Earth, the perspective has been very productive. Especially as we have explored the smallest of objects on Earth down to the quantum scale it has worked well. However, at the scale larger than Earth, it begins to fail; the larger the scale the more consistently and the more completely it fails. At the cosmic scale, it fails to explain the nature of the growth of the universe, which is why a patch called "Dark Energy", is added. At, the somewhat smaller scale of galaxies, a particle based perspective fails to explain how galaxies form and hold together; which is why a patch called "Dark Matter, has been invented. At the even small scale of black holes, the particle based perspective leaves fundamental gaps of understanding. Even at the smaller scale of our sun, there are phenomena that a particle perspective fundamentally fails to explain.

The conclusion to draw from these facts is that a bottom up view of a world built of particles is valid, but not a universe built bottom up of particles. The Higgs boson provides a bridge between our particle perspective and an omnipresent fluid perspective, but the bridge only helps if we cross it. According to the particle perspective, the Higgs boson is a particle associated with an infrastructure field that underlies the universe. It says that the field exists, but to cross the bridge we must balance the emphasis on particles with an emphasis on the field, i.e. the fluid, as the fundamental elemental building block of the universe. Then having crossed that bridge, we keep the bridge. We use both the particle perspective and the hyperfluid perspective in their appropriate regimes.

Whereas the particle perspective works well for small scale physics, the hypervortex model explains the cosmic scale phenomena from a top-down perspective. The hyperfluid view says that the universe expands because of overall forces on the hyperfluid. That expansion causes turbulence that causes large scale vortices such as galaxies. The galaxies spin off smaller vortices such as black holes and solar systems. The black hole at the center of the galaxy is the center of the vortex that is the galaxy. And in all of that, vortices of the smallest particles are also created. Then the small vortices build up into bigger things. For objects between the smallest and the largest, their structure and behavior results from a combination of the top-down hyperfluid effects and the bottom-up forces among the smallest particles. The relative mix of fluid and particle effects probably shifts more to the fluid at the large scale and more to the particle at the smaller scale.

Thus, this one section has introduced a perspective that is a synthesis of two traditionally opposing perspectives. While the universe is fundamentally hyperfluid, that gives rise to particles. Then, the effects that we think of as fluid dynamics and the effects that we think of as particle dynamics both affect various objects in various degrees depending on the size of the object. Now this may seem to be stated as a nebulous postulate, it is not a postulate at all. This may sound comparable to Einstein's statement of wave-particle duality, and if the next chapter did not have the equations that quantify the implications of this synthesis, it would be comparably ambiguous to a statement of wave particle duality. However, the next chapter does lay out the equations and it does derive the synthesis. It derives for example, the relation between electromagnetic fields and the electric charge on particles. Also, it derives how fluid behavior gives rise to gravity and hence mass. This section is not postulate; it is a commentary stating that one does not have to discard the particle perspective, but rather one must synthesize such perspective with the hyperfluid perspective.

8. Derivation and Mathematical Rigor

This chapter provides the mathematical rigor and details that make the preceding chapters a description of a model, rather than speculation. For brevity, the chapter focuses on aspects of unification that have been the most challenging for other models. Also for brevity, this chapter assumes substantially more knowledge of mathematics and physics on the part of the reader than is assumed in other chapters.

The focus here is to derive the Lagrange equation that unifies physics. From the Lagrange equation it derives correct equations for electromagnetic forces including some details beyond what other models can derive. It further derives equations for gravity, including a derivation showing that inertial mass and gravitational mass are the same. Additionally it shows how gravity and electromagnetic forces couple and how that coupling generates quantum mechanics. A basic construction of Special Relativity in a Euclidean hyperspace, previously published in Reference 1, is not repeated here. Rather, this chapter summarizes and generalizes the salient points of that work, especially the point that Special Relativity can occur in a Euclidean hyperverse. This chapter then uses that work to build the unifying Lagrange Equation.

Because the model's mathematics is constructed in a flat-space, its mathematical complexity is reduced as compared to most discussions of unified physics. Trigonometry is sufficient for some of this chapter's discussions. Because of the relative simplicity of the mathematics, mathematical notations from general relativity are not used. For those familiar with such notations, be careful to recognize that the notation here is more that of undergraduate mathematics such as that used in linear algebra.

The mathematics here is less complex than mathematics used to predict Earth's weather. The hyperfluid has fewer attributes than air and therefore fewer equations. The hyperfluid has equations only for density and velocity, whereas weather modeling has equations at least for temperature, pressure, density, composition, entropy, and velocity.

Still, just as numerical methods are used to compute the weather; numerical methods will likely be useful to analyze hypervortices.

The Coordinate System: The coordinate system used here is a Euclidean spatial coordinate system with four spatial coordinates. The coordinates can be labelled as (x, y, z, w), where (x, y, z) are the familiar 3 spatial coordinates that we observe and w is the additional coordinate that we do not observe. Here x_μ is also used, where μ takes on the values 1, 2, 3, 4 to refer to the individual coordinates (x, y, z, w), respectively. Also, x_i refers to the coordinates (x, y, z), omitting the unobservable coordinate.

As in any Euclidean coordinate system, the distance between two events in the hypervortex model depends on their location as:

$$ds^2_{Euclidean} = dx^2_\mu$$

In the notation throughout this chapter, $dx^2_\mu \equiv dx_\mu dx_\mu$ is the dot product of dx_μ with itself. Also, the same notation is used for various vectors and vector fields in the chapter. Similarly for any two vectors, $X_\mu Y_\mu = X_1 Y_1 + X_2 Y_2 + X_3 Y_3 + X_4 Y_4$, i.e., the dot product of X_μ with Y_μ.

In contrast to the Euclidean definition of distance, Einstein's mathematics for Special Relativity uses Lorentz coordinates, where the distance between two events is:

$$ds^2_{Lorentz} = dx^2_i - c^2 dt^2$$

In the above, and throughout, c is the speed of light and t is time. Special relativity in Lorentz coordinates mixes space and time in the sense that the time between events affects the distance between the events. Such is not so in the hypervortex model. In the hypervortex model, the distance between events depends only on their spatial location. Time does not affect that distance. In the hypervortex model, time is separate from space, just as it is in our everyday lives.

Since the onset of the use of Lorentz coordinates, theorists have tried to make space more Euclidean by defining a coordinate $w = ict$. Here, "i" is the standard imaginary number i, suggesting that w is an imaginary coordinate. With such a definition, the Lorentz equation for distance can be written to look exactly like the Euclidean definition of distance. However, those theorists were unwilling or unable to make the next leap – to declare w as real coordinate, and then to explore the properties of w that give rise to our observed universe. Such a leap is an essential element of the work here and is explored in upcoming sections of the chapter.

Causality: In general, causality examines whether, given two events, the first to occur can cause the second to occur. In general, if the two events are some distance apart, then the first may cause the second if the effects of the first can propagate to the location of the second in time for the second to occur. If the effects of the first event cannot propagate that fast then the first event cannot cause the second event. Thus, in general, causality depends on the speed of propagation of effects, which may depend on the mechanism of propagation.

In Special Relativity the Lorentz distance is said to determine whether two events may be causally related. If the distance is negative, then they are causally related and if the distance is positive then they are not. The Lorentz distance applies the general concept of causality but specifically using the speed of light as the maximum rate at which the effects of the first event propagate to the second. That is, it couples causality specifically to the speed of light and makes that coupling specific to the nature of space and time. Consequently, Special Relativity, says that light is the fastest propagating thing in the universe.

"Spooky Action" disproves the notion that the Lorentz distance determines causality, which was Einstein's point in using the adjective "Spooky". The mathematics of Special Relativity makes the Lorentz distance the determiner of causality, and ergo, the mathematics of Special Relativity is wrong. Stated more strongly, the experiment that proved the existence of "Spooky Action" really proved that something propagates faster than the speed of light and since Special Relativity predicts that the speed of light is the maximum propagation speed, hence Special Relativity is wrong. These were Einstein's interpretations, which others have ignored. The hypervortex model accepts Einstein's interpretation, removes the major culprit – the Lorentz distance – by defining w as real, and re-derives the physics of Special Relativity using Euclidean distance.

Clocks and Action: In Special Relativity the rate of time is locked to the rate of clocks. Wherever one is, the rate of the local clock is the local rate of time declares Special Relativity. Such a definition of time is a choice. There is an alternate choice.

In actual practice, clocks repeat an action. We count the repetitions of that action and use that to infer a time. Mathematical physics provides a general equation regarding the Action performed by any system, including clocks:

$$Action = \int L dt dx_\mu$$

In the equation, L is called the Lagrangian density of the system after the equation's author, Lagrange, and it describes the properties of the system, e.g. the hyperfluid. The equation shows that the Action of a system depends on both the properties of the system and the properties of space and time itself, represented above by $dt dx_\mu$.

The choice made by Special Relativity declares L to be constant, and, therefore, all the complexity of the equation becomes assigned to space and time in the equation. This choice is the root cause for the need for complex non-Euclidean mathematics in Special and General Relativity.

The hypervortex model makes the opposite choice. It declares space and time to have a simple constant nature, and puts the complexity into L, that is, it puts the complexity in the properties of stuff in space rather into space and time itself. This alternate choice removes the Lorentz definition of distance, which removes the "Spooky" nature of faster than light coupling between events. Further since particle physics also puts the complexity into L and not into space and time, this alternate choice starts to unify our mathematics for particle and fluid aspects of the stuff in space.

In the hypervortex model, there is a conversion between the clock rate and a global time. Global time, t, is the same everywhere. Clock rates vary and knowledge of the Lagrangian enables accurate conversion between the clock rate and global time. Such conversions are used in practice worldwide, across the solar system, wherever man has placed clocks that require high accuracy.

In this work "local time" refers to an uncorrected time as read directly from a local clock. The rate of aging has been shown for some systems to match local time. Determining whether this is true for all systems is beyond scope for the book.

Lagrange's derivation of Action showed that any coordinate system that can represent the full scope of a system's behaviors is acceptable. Thus, choosing Euclidean coordinates is acceptable, as long as enough spatial dimensions are included to span the actual dimensionality of space. The chapter does not attempt a general proof of this. Rather, it assembles a Lagrangian density and shows that its properties provide a framework for a unified hypervortex model.

Relativistic Motion in Euclidean Hyperspace: Attempts have been made to explore Special Relativity as a hyperspace in which $w = ict$. A search on the World Wide Web for the term "world line" finds various such efforts. However, those world-line constructs, in general, violate the principle of relativity in many ways and thus do not succeed. In the world-line concept, world-lines are assigned certain behavioral similarities to the vortices of the hypervortex model. In particular, each world-line extends one particle into hyperspace and the observed particle is at the intersection of the world-line and the observable 3D universe. However, world-lines are necessarily stationary in hyperspace because of the fixed relation between w and t as $w = ict$. The world-line cannot move over time. Only the observable 3D universe can move and it moves along w at the nominal rate c. The problem with the world-line concept is that there is no motion of the 3D universe that could reproduce Einstein's equations for Special Relativity while obeying the principle of relativity everywhere. In such models there is no possibility of allowing the world-lines to move, because that would violate the fixed relation between w and t as $w = ict$.

The hypervortex model solves the problem by declaring w real and distinct from time. The hyperfluid's vortices replace the world-lines of the other concepts. The vortices are allowed to move over time because space and time are decoupled by the addition of the w coordinate. Particles are observed at the intersections of moving vortices and moving 3D universes. The ability of vortices to move over time allows successful construction of Special Relativity while obeying the principle of relativity everywhere in the Euclidean Hyperverse. Reference 1 provides one derivation of such equations. Here a generalization of that approach is provided. Such generalization is important to allow various readers to explore model variants that might improve on the model provided here.

The first step to reproducing Special Relativity in a Euclidean Hyperverse is to reproduce Einstein's equations for "relativistic mass" and "time dilation". To do so, for simplicity, generally think of the (x, y, z) coordinates as being oriented so that locally (at the laboratory where observations are being made) they are the three observable dimensions of the local observable universe, while the new coordinate, w, is generally oriented perpendicular to the observable universe. Further, consider the origin of the coordinate system as being set such that our observable universe is, on average, at $w = 0$. With this orientation, other parallel observable universes are, approximately, specifiable by referencing locations at $w \neq 0$.

Many calculations can be further simplified, without loss of generality, by considering a particle moving along coordinate x at some velocity, v. Then the y and z coordinates can be ignored when analyzing simple inertial motion, and we can construct equations for the overall motion of a vortex such that the observed particles have relativistic mass and time dilation. The remainder of this section uses this simplification.

In the hyperfluid, relativistic mass of observed particles is determined by two physical factors: (1) the mass per unit length of the vortex, and (2) the length of vortex that lies within the observable universe. The product of these two factors is the observed mass. There are multiple ways that the two factors can vary such that their product matches Einstein's result for relativistic mass. A typical approach from historical efforts using world-lines is to assume that the mass per unit length of vortex is constant, which makes sense physically. Thus, the observation of increasing mass as velocity rises becomes a result of an increase in the length of vortex that is within the observable universe. If an observable universe has a finite thickness, ξ, then the length of vortex in an observable universe is determined by the angle that the vortex makes with the observable universe. The observed mass of a particle is its rest mass, m_0, when the vortex is perpendicular to the observable universe. m_0 is the product of ξ and the mass per unit length. The observed mass rises as the angle changes to be less perpendicular. The hypervortex model reproduces Einstein's equation for relativistic mass by setting an appropriate relation between velocity and the angle of the vortex to the ambient 3D universe.

To reproduce time dilation, the clock rate of a particle is related to the motion of the vortex. Again leveraging the concepts from world-line models, the observed clock rate becomes a product of two factors: (1) the length of vortex that passes through the observable universe per unit time, and (2) the amount of observed Action that occurs per length of vortex that passes through the observable universe. Following the assumptions used from historic world-line attempts at a hyperverse model, the length of vortex that passes through the observable universe per unit time is assumed to be independent of the observed velocity. That is, no matter what the velocity of the particle and thus no matter what the angle of the vortex relative to the observable universe, the same length of vortex passes through the observable universe per unit time. Thus, the hypervortex model reproduces Einstein's equations for time dilation by setting the relation between the velocity of the vortex along the w coordinate and the observed velocity, v, of the observed particle such that the length of vortex passing through the observable universe is the same no matter the observed velocity of the particle.

One particular set of equations for relativistic mass and for time dilation in the hyperfluid is quite simple. First, the observed speed of a particle is given by $v = c sin(\theta)$, where θ is the angle of the vortex relative to w. The observed relativistic mass is $m = m_0/cos(\theta)$; this equation is recognizable as being the same as in the standard relativity equations when one realizes that

$$\frac{m_0}{cos(\theta)} = m_0\left(1 - sin^2(\theta)\right)^{-\frac{1}{2}} = mc\left(1 - \frac{v^2}{c^2}\right)^{-\frac{1}{2}} = \gamma mc$$

Here γ has the same meaning as in Einstein's formulation. The observed momentum is $mc tan(\theta)$. Vortices have variable speed along the w direction; that speed is $c cos(\theta)$. Whether the vortex moves in the positive or negative direction along the w coordinate is dependent on details. In addition to reproducing relativistic mass and time dilation, a good model must also reproduce other aspects of Einstein's Special Relativity such as: (1) the principle of relativity itself and (2) addition of relativistic velocities. Reproducing the principle of relativity is important for constructing the model and is discussed in the next section. The addition of relativistic velocities amounts to doing a particular check that the equations work; the check has been made, but is not included here.

Satisfying Relativity in Euclidean Hyperspace: In the hyperfluid model, our measurement tools and the speed of light are affected by their location in space. Our measurement tools are clocks and rulers, and often our ruler is the distance travelled by light in a given time. The rate of clocks varies in space and time. The symbol used here to represent the local clock rate is t_L. The rate, t_L, varies in space and time, which in the parlance of mathematical symbolism is stated as $t_L(x,t)$. Similarly, the speed of light, c, and the length of rulers, r, vary in space and time, which, using the same parlance is denoted as $c(x,t)$, and $r(x,t)$.

The principle of relativity will be satisfied if it is impossible for an observer to distinguish one place in the universe from another by local measurement. To achieve the principle of relativity, rulers, light speed, and clock rates change in such a way that we do not observe those changes by local measurement. Mathematically, that is represented as $ct_L/r = 1$ or $ct_L = r$. That is, the speed of light times the clock rate is equal to the length of the ruler. At any location in space at any given time, there is one local clock rate, one local speed of light, and one local ruler scale. After all, these are physical devices and we cannot physically place two clocks or two rulers in one place at one time.

While the above may seem to imply that it is impossible to observe our measurement tools changing, experiments have observed these changes. For example, such changes are observed for clocks via non-local measurement. A clock measures the cumulative effects of the history of its placement. Thus, put it somewhere where clocks run slow (or fast) for a while and bring it back and one can see that the clock ran slow for a while. This has been done and it works. The same cannot be done for light or rulers. They change and change back when moved somewhere and returned.

To achieve the principle of relativity requires more than a rescaling of our measurement tools as velocity changes. Consider our observation of a satellite moving inertially along x at a high velocity v. Time dilation slows the clock in the satellite by a factor of $1/cos(\theta)$ compared to a clock on Earth. This means that a local observer, in the satellite, takes longer to declare that a second has elapsed than an observer watching a clock on Earth. That further means, assuming the speed of light is the same at the satellite as on Earth, that, in the time it takes to observe one second pass in the satellite, light has travelled farther by a factor of $1/cos(\theta)$ than it does on Earth. In order for the observer in the satellite not to notice this difference, the ruler in the satellite must stretch also by a factor of $1/cos(\theta)$ compared to its size on Earth. Indeed everything near the satellite must stretch by this factor in all directions, including the vortices of all the particles comprising the satellite. In order for everything to stretch in all directions, a deformation of the fluid occurs.

The deformation of the fluid travels with the vortex; the vortex causes the deformation. The deformation has three aspects: the time dilation, the spatial stretch and a rotation. Specifically, near the satellite the expansion flow rotates to remain parallel to the vortex of the satellite, just as it was parallel to the satellite's vortex when the satellite was at rest on the Earth before launch. Thus, the vortex angle, θ, described previously, is an angle relative to the ambient expansion flow, not relative to the local expansion flow at the satellite. That is, the local observable universe also rotates by the angle, θ, so that it is perpendicular to both the local expansion flow at the satellite and to the satellite's vortex.

Rotation of the expansion flow and the corresponding rotation of the observable universe together enable the local volume of the observable universe to stretch by the factor $1/cos(\theta)$, as required to implement the principle of relativity. This rotation also puts any local observer into a reference frame that is at rest relative to the satellite (or whatever object is in motion), as required to implement the principle of relativity. The rotation, time dilation, and expansion of the observable universe, all centered at the location where the object's vortex intersects the observable universe, implement Einstein's equations of Special Relativity. This implementation, however, is a physical change to the hyperfluid not a transformation of space itself.

Density of the Hyperfluid: Hyperfluid is considered in this model as the underlying matter in a hyperverse. The amount of this matter can vary throughout a hyperverse. Because it varies, a mathematical symbol, a variable, is used to represent it. Here the amount of that matter at each location in a hyperverse is represented as M. At times, to explicitly show that its value varies across space and time, it is represented as $M(x_\mu, t)$. $M(x_\mu, t)$ is called a field because that is the standard generic name of properties that vary throughout a space. It is called a scalar field because it has only one value at each location in space at any time. Technically, it is the density of the matter at each location in a hyperverse; the total amount of hyperfluid matter in a hyperverse is the integral of $M(x_\mu, t)$ over the entire hyperverse.

The symbol "M" is used here for the density of the hyperfluid for two reasons. First it gives some of the following equations a familiar form. Second, it is likely that the hyperfluid density will be found to be a mass density. Note that the density of the hyperfluid is represented by an uppercase "M". This is to distinguish it from the masses of particles; particle masses are represented by a lowercase "m".

Velocity of the Hyperfluid: The velocity of the hyperfluid may vary throughout a hyperverse. The velocity is a vector field because at every location it has four values; each value is a speed along one dimension – four dimensions imply four values. If there are additional dimensions, then the velocity has more values at each location. The symbol used here to represent the velocity field is V_μ. V is the field name, and the subscript μ is included to indicate that it is a vector field with a component for each dimension. μ takes on different values to reference the velocity along one specific direction. Sometimes μ may be replaced with a specific coordinate symbol to reference a particular coordinate's velocity. Thus V_R is the velocity along the R coordinate (when using spherical coordinates). The fields M and V_μ are the only two fields used at this time to represent physical features of the hyperfluid. The hyperfluid has no entropy, no temperature, or other internal property. The fluid is not made of two components; it has one component with one density and one velocity.

Momentum of the Hyperfluid: Momentum in the hyperfluid is defined as an additional vector field for convenience of the mathematics; it is not an independent field from density and velocity. The symbol P_μ is used here for the momentum. At every location in a hyperverse, the matter underlying the hyperfluid has a density and it has a velocity. Their product is the momentum of the fluid at that location. Thus, the computation of the momentum is $P_\mu = MV_\mu$. Technically, P_μ is a momentum density, momentum per unit volume of a hyperverse.

Kinetic Energy of the Hyperfluid: The kinetic energy of the hyperfluid is the kinetic energy of a hyperverse. A hyperverse's kinetic energy is not fluid energy plus particle energy. The particle energy is the portion of fluid energy contained in vortices in the fluid. As with any fluid, its kinetic energy is the sum of the kinetic energy of all of the fluid everywhere in a hyperverse. In accord with the computation of the kinetic energy for ideal fluid models in general, at each location in a hyperverse, the kinetic energy is given by $K.E. = \frac{1}{2}MV^2$; that is, it is the amount of fluid at that location times $\frac{1}{2}$ the square of its velocity at that location. The total kinetic energy of the fluid, then, is the sum over all locations in a hyperverse of the kinetic energy at each location. There is no relativistic term in the kinetic energy. The kinetic energy can equivalently be written in terms of fluid density and fluid momentum as $K.E. = P_\mu^2/2M$.

Potential Energy of the Hyperfluid: Potential energy in an ideal fluid is energy stored in the variations in the density and velocity of the fluid. As potential energy, it is convertible to kinetic energy, and back, losslessly. The sum of the kinetic energy and the potential energy is a constant summed over all the fluid. The form of the potential energy determines detailed behaviors of the fluid, which in turn determine the shapes and sizes of vortices, the speeds of various waves, and the observable forces among the vortices and waves. A key area of ongoing research is to determine the correct form of the fluid's potential energy such that its predictions match all measured observables. This section selects a subset of the possible potential energy terms to exemplify how known observable physics is caused by properties of the hyperfluid. Variant and additional terms may provide better results later.

Because the universe is observed as a generally continuous system, where discontinuities in a field's values are rare or non-existent, potential energy terms that prevent discontinuities are appropriate. Thus, potential energy terms are preferred whose values increase as gradients of the fields increase. Also, potential energy terms that have a quadratic form, i.e., that depends on the square of a gradient, have the nice physical property (and generally observed physical property) that the process of summing energy components over space is commutative. That is, the total energy is the same whether summed first over each dimension and then over space, or summed over dimensions and then over space. Further, the equipartition theorem provides useful information about how the total energy is partitioned among various quadratic energy terms. Moreover, such mathematical properties of quadratic energy terms are consistent with the known physical properties of space. Thus, in this model quadratic potential energy components have been given priority and since their use has provided successful results, exploration of alternatives has been neither required nor pursued.

A potential energy term used in Reference 1 is $k_M/2 \left(\partial M / \partial x_\mu \right)^2$, where k_M represented a constant stiffness (inverse of the compressibility) of the fluid and the term in parenthesis meant the gradient of M. The gradient of a field, represented as $\partial / \partial x_\mu$, is a standard mathematical function. When applied to M, $\partial / \partial x_\mu$ provides rate of change of M along each dimension. It can also be viewed as providing the maximum rate of change of M at that location and the direction in which that rate of change occurs. Both views are equivalent and that knowledge is standard knowledge that is basic to engineering mathematics. The gradient of a scalar function is a vector function. That is, $\partial M / \partial x_\mu$ at each location is a vector whose components provide the rate of change of M along each direction (w, x, y, z). The expression $\left(\partial M / \partial x_\mu \right)^2$ means that the square of each component of the gradient is computed separately and then summed.

In this work, the potential energy term used for gradients of the fluid density is slightly generalized relative to the prior work. In particular, the generalization allows the stiffness to vary as the density varies. The allowance that the stiffness varies as a function of the fluid density is shown in the equations as $k_M(M)$. Additional potential energy terms that depend on the fluid density can be added, such as M^2 and $(\partial M / \partial t)^2$. Their addition would not be a fundamental change to the model. They are not added here as there has been no reason to do so.

A potential energy term related to gradients in the velocity is also added in this work. The new energy term limits transverse gradients in the velocity, as a viscosity would, but without loss of energy, i.e., without violating Newton's Laws. Without this term, in what is perhaps a quirk of the mathematics, the gradients of the velocity will still be continuous. Indeed, without addition of the new energy term, the fluid will always have zero vorticity, even while allowing the existence of vortices. For these reasons, the previously published work (Ref. 1) omits such a potential energy term. However, it is added here because allowing vorticity in the fluid leads to presence of electromagnetic behavior in the fluid.

The form of the potential energy term used here that involves velocity gradients is $\left(\partial P_\mu/\partial x_\nu - \partial P_\nu/\partial x_\mu\right)^2$. The subscript ν, like μ, represents a spatial dimension. For the selected potential energy term, there is a potential energy component for each combination (μ, ν). There are thus 16 quadratic energy terms in four spatial dimensions. However, the terms are zero for $\mu = \nu$. Also, pairs of terms that have μ and ν swapped have the same energy. Thus there are only six distinct terms. The terms also include a coefficient $k_V/2$. Also, the coefficient here has the same value for all 16 quadratic terms components, because the universe is observed as being isotropic (independent of direction).

The derivation currently assumes that k_V is a constant. Dependencies on the fields (M and P_μ) can be added later without fundamental change to the model. The potential energy term is written in terms of the momentum rather than the velocity because it simplifies the model mathematically and seems to map well to observed physics. Examples of other forms that can be used in addition, or as replacements, include $\left(\partial P_\mu/\partial x_\mu\right)^2$, $\left(\partial V_\mu/\partial x_\nu - \partial V_\nu/\partial x_\mu\right)^2$, $\left(\partial V_\nu/\partial x_\mu\right)^2$, $\left(\partial P_\nu/\partial x_\mu\right)^2$, and $(\partial P_\nu/\partial t)^2$. Each such term may be added with its own multiplier indicating the relative amplitude of that part of the total potential energy.

Continuity in the Fluid: Except perhaps at some special locations in a hyperverse, the hyperfluid is postulated to be conserved. Thus, it cannot simply disappear in one place and reappear in another.

There is a standard equation in physics to represent this constraint. Using the fields defined above, the continuity constraint is written as

$$\partial M/\partial t + \partial P_\mu/\partial x_\mu = 0 \qquad \binom{Continuity}{Equation}$$

While this continuity equation looks complicated it simply says that the amount of fluid entering or leaving a region in space is equal to the change in the amount of fluid in that region of space. The term $\partial P_\mu/\partial x_\mu$ is the divergence of the momentum, or, equivalently, the amount of flow of fluid leaving each region in space. The term $\partial M/\partial t$ is the rate of change of the density of the fluid due to that flow. The equation applies at all locations in space at all times, at least until such time as a fluid source or sink is found in the real world. Such sources and sinks would not invalidate the model. Rather, when and if such phenomena are found, they refine the model by refining this constraint.

The Conservation Field: In order that the equations of motion for the hyperfluid satisfy the continuity equation, the continuity equation must be combined with all the other terms that determine the equations of motion. Fortunately, the Lagrange mathematics that will be used shortly to determine the equations of motion from all of the above parts provides a simple means to do this. It requires the addition of a new field describing this aspect of the hyperfluid. The symbol used here for the new field is λ. It is a scalar field that may vary by location and time. Thus, at times the field may be shown as $\lambda(x_\mu, t)$. The field $\lambda(x_\mu, t)$ is a special type of field called a Lagrange multiplier, and λ is the commonly used symbol to represent such fields. This field is not a physical property of the hyperfluid though at many times that subtlety may be irrelevant. Its dependence on location and time is determined by solving the equations of motion together with the continuity equation. The conservation field does not contribute to the energy of the hyperfluid, or to its density. It is a manifestation of the property of conservation of hyperfluid, and its value over space and time indicates the relative impact of that property on the motion of the hyperfluid.

Lagrangian for the Hyperfluid: The Lagrangian density is assembled according to a simple standard formulation. In particular it is the kinetic energy, minus the potential energy, plus terms that specify constraints on the values of the variables and fields. The Lagrange density, L, assembled from the components described above, is:

$$L = \frac{P_\mu^2}{2M} - \frac{k_V}{2}\left(\frac{\partial P_\mu}{\partial x_\nu} - \frac{\partial P_\nu}{\partial x_\mu}\right)^2 - \frac{k_M(M)}{2}\left(\frac{\partial M}{\partial x_\mu}\right)^2 + \lambda\left(\frac{\partial M}{\partial t} + \frac{\partial P_\mu}{\partial x_\mu}\right)$$

Lagrange Density Equation: One equation for everything in a hyperverse.

The above Lagrangian density differs from the form on the book cover and in the recap; those versions resolve the M dependence of $k_M(M)$, which is derived later in the chapter.

The first two terms in the Lagrangian density relate to electromagnetism. The third term adds gravity. The fourth term adds the detailed coupling between electromagnetism and gravity. Also it determines the set of elementary particles and their internal structure, and it adds the strong and weak forces and quantum mechanics.

The Lagrangian satisfies the principle of relativity. That is, mathematically, any change to the origin of the coordinate system or to the velocity of the coordinate system, does not affect the potential energy terms in the Lagrangian. For example, the gradients of the momentum are transverse to the velocity so that any added constant drops out in the potential energy.

Equations of Motion: The equations of motion for everything in the fluid come from the Lagrangian and from the continuity equation. To enable solving the equations of motion, it is required that there be equal number of equations as the number of fields, which there are. To get the equations of motion, the Lagrange differential is computed for each of the fluid's fields. For any field f the generation of the equations of motion from the Lagrangian is standard mathematics. The following mathematical operation creates a set of equations that are solved to determine the detailed behavior of the hyperfluid, and hence of a hyperverse and of the observable universe.

$$\frac{\partial L}{\partial f} - \frac{\partial}{\partial t}\left(\frac{\partial L}{\partial \frac{\partial f}{\partial t}}\right) - \frac{\partial}{\partial x_\mu}\left(\frac{\partial L}{\partial \frac{\partial f}{\partial x_\mu}}\right) = 0$$

Standard equation for generating equations of motion from a Lagrangian density.

The equations of motion for the hyperfluid are obtained by applying the above operation, substituting the matter density M and momentum P_μ one at a time in place of f. The above operation is not applied for the conservation field λ. The density equation below results from substituting M. The momentum equations below are the result of substituting P_μ. The third equation below is again the continuity equation restated here for completeness.

$$-\frac{P_\mu^2}{2M^2} + k_M'(M)\left(\frac{\partial M}{\partial x_\mu}\right)^2 + k_M(M)\frac{\partial^2 M}{\partial x_\mu^2} - \frac{\partial \lambda}{\partial t} = 0 \qquad \left(\begin{array}{c}Density\\Equation\end{array}\right)$$

$$\frac{P_\mu}{M} + 2k_V\frac{\partial}{\partial x_\upsilon}\left(\frac{\partial P_\mu}{\partial x_\upsilon} - \frac{\partial P_\upsilon}{\partial x_\mu}\right) - \frac{\partial \lambda}{\partial x_\mu} = 0 \qquad \left(\begin{array}{c}Momentum\\Equations\end{array}\right)$$

$$\frac{\partial M}{\partial t} + \frac{\partial P_\mu}{\partial x_\mu} = 0 \qquad \left(\begin{array}{c}Continuity\\Equation\end{array}\right)$$

Unified equations of motion.

In the above, the term $k_M'(M)$ is the derivative of the stiffness of the hyperfluid with respect to M. There is one density equation. In it, there are terms for each value of μ. The Momentum Equations have an equation for each value of μ, in accord with the prior discussion regarding the notation. Each Momentum Equation includes terms for each value of υ. There is one continuity equation. In it, there are terms for each value of μ.

The three above equations determine everything about the universe. The momentum equation and the continuity equation turn out to reproduce Maxwell's equations for electromagnetism. That derivation is presented later in the chapter and includes relating the value of the constant, k_V, to the overall scale of the matter density, M and the momentum, P_μ. The density equation is the truly new equation. It turns out to unify gravity with Maxwell's equations. A derivation later in the chapter shows that $k_M(M) = k_G/M$ and thus $k'_M(M) = -k_G/M^2$, where k_G is a constant. Determination of the value of, k_G and of the overall scale of the density, M, flow from that derivation, but those details are left for the future. Aspects of the momentum and continuity equations couple them to the density equation. That coupling integrates the strong and weak forces and Quantum Mechanics.

Structure in the Equations of Motion: Simple rearrangement of the Density and Momentum equations exposes structure in the equations. The rearranged structure puts the terms that couple the density and momentum equations to the right of the equal sign, as follows:

$$k'_M(M)\left(\frac{\partial M}{\partial x_\mu}\right)^2 + k_M(M)\frac{\partial^2 M}{\partial x_\mu^2} = \frac{\partial \lambda}{\partial t} + \frac{P_\mu^2}{2M^2} \qquad \begin{pmatrix} Structured \\ Density \\ Equation \end{pmatrix}$$

$$2k_V\frac{\partial}{\partial x_v}\left(\frac{\partial P_\mu}{\partial x_v} - \frac{\partial P_v}{\partial x_\mu}\right) = \frac{\partial \lambda}{\partial x_\mu} - \frac{P_\mu}{M} \qquad \begin{pmatrix} Structured \\ Momentum \\ Equations \end{pmatrix}$$

Structured equations of motion.

In the rearrangement, particles and structures in a hyperverse are represented by the terms on the right in the equations, while the terms to the left represent the response of the ambient hyperfluid to the behavior of the particles on the right. The terms on the left in the Structured Density Equation relate to gravity and quantum mechanics. The terms on the left in the Structured Momentum Equations equate to Maxwell's Equations for electromagnetic systems.

The equations of motion are used to determine M and P_μ for any scenario of interest, thus determining the fluid density and momentum at every location in space and over time. Various methods may be used to solve the equations for M and P_μ. By setting the right sides of the above equations to zero, the left sides of the equations can be solved by themselves to provide static solutions, i.e., solutions in which the fluid density and momentum at each location in space do not change over time. When static solutions do not exist, temporal evolution of the fluid occurs via coupling to the right sides of the equations. The terms on the right include explicit change over time. These changes cause the terms on the left to change. The terms on the left again seek a stationary solution in response. Those changes generate forces that induce more change to the terms on the right. Applied in sequence, the behavior of a hyperverse unfolds over time. Observers in observable universes experience a timeline that combines two effects: (a) real temporal evolution of the whole of a hyperverse, and (b) motion of the observable universe through a hyperverse.

The next few sections study the equations in special cases when the coupling among the equations and the complexity of their solution is reduced. Then, in subsequent sections, the coupling and complexity is reintroduced to study additional aspects of the equations.

Covariant Electromagnetism: This section shows that the second term in the Lagrangian density corresponds to the electromagnetic forces. It also shows that the equations of electromagnetism are exact only in the limit when the density, M, of the fluid is constant over time.

First, the divergence of P_μ in the continuity equation equals zero if and only if the density, M, of the fluid is constant over time. This property of P_μ can be satisfied by the assignment $P_\mu = k_A A_\mu$ where k_A is a constant that depends upon the choice of units, [of which there are several common ones], and where A_μ is the 4-vector potential used in covariant formulation of electromagnetism. (See, for example Reference 3.)

To now identify electromagnetism within the hyperfluid Lagrangian density, first a quick review of covariant formulation of electromagnetism is provided. Per Reference 3, using Gaussian Units, a variable $F_{\mu\nu}$ is defined for convenience as

$$F_{\mu\nu} \equiv \left(\frac{\partial A_\mu}{\partial x_\nu} - \frac{\partial A_\nu}{\partial x_\mu} \right)$$

This definition allows writing the relationship between A_μ and the electric 4-vector electric current, J_μ, as

$$\frac{\partial F_{\mu\nu}}{\partial x_\nu} = \frac{4\pi}{c} J_\mu$$

Also, it allows writing the electromagnetic energy as

$$Energy\ Density\ (per\ m^3) = \frac{1}{16\pi} F_{\mu\nu} F_{\mu\nu}$$

Now, to reformulate the above in terms of P_μ rather than A_μ for the hypervortex model, define

$$G_{\mu\nu} \equiv \left(\frac{\partial P_\mu}{\partial x_\nu} - \frac{\partial P_\nu}{\partial x_\mu} \right) = k_A F_{\mu\nu}$$

This allows writing equations for J_μ as

$$\frac{\partial G_{\mu\nu}}{\partial x_\nu} = \frac{4\pi k_A}{c} J_\mu \qquad \left(\begin{array}{c} Relation \\ Between\ J\ and\ G \end{array} \right)$$

Similarly, the corresponding energy density can be written as

$$Electromagnetic\ Energy\ Density\ (per\ m^3) = \frac{1}{16\pi k_A^2} G_{\mu\nu} G_{\mu\nu}$$

So far, the above Electromagnetic Energy Density, despite the different appearance, has the same dependence on M and P_μ as the second term in the Lagrangian density, the term containing k_V. However, the Lagrangian density is written in terms of energy per m^4 while the above energy density for electromagnetism is per m^3. The difference is that electromagnetic energy has always been implicitly integrated over the thickness of one observable universe. Thus, to get electromagnetic energy per m^4 the above Electromagnetic Energy Density is divided by the thickness, ξ, of an observable universe. This adjustment gives electromagnetic energy density as

$$Electromagnetic\ Energy\ Density\ (per\ m^4)\ =\ \frac{1}{16\pi\xi k_A^2}G_{\mu\upsilon}G_{\mu\upsilon}$$

Thus, the equation for electromagnetic energy from Maxwell's Equations will match the second term in the Lagrangian density by setting

$$k_V \equiv \frac{1}{16\pi\xi k_A^2} \qquad \begin{pmatrix} Electromagnetics \\ Normalization \\ Relation \end{pmatrix}$$

The values of ξ and of k_A are both unresolved at this point so that above equation does not set a numeric value to k_V. It has been considered that $k_A^2 = \mu_0$ for the equations written in some choice of units. Work to prove a relation between k_A and μ_0 is ongoing.

The relations derived in this section for J_μ and k_V can be used in the Structured Momentum Equations to provide a new equation relating the electric current to the structure terms in the equations of motion as follows:

$$\frac{\partial\lambda}{\partial x_\mu} - \frac{P_\mu}{M} = \left(\frac{\partial\lambda}{\partial x_\mu} - V_\mu\right) = \frac{1}{2\xi k_A c}J_\mu \qquad \begin{pmatrix} Charge \\ Current \\ Equations \end{pmatrix}$$

The above provides new information about electric charge. Whereas typical models consider charge as a distinctly separate aspect of nature, the above equation shows that electric charge is our observation of a particular behavior of the hyperfluid – a behavior that occurs due to the continuity constraint's effects on hyperfluid motion.

Notice that electromagnetism, as formulated in the second term of the Lagrangian, has no explicit time dependence. That is, t, the variable for time does not appear in that term of the Lagrangian. Also, notice that none of the four spatial coordinates is special in the formulation; the equations are the same for all four spatial coordinates. The symmetry is broken and time is introduced by the expansion current. To show this, the next few sections develop the Lagrangian's implications for the expansion current, which then allow further discussion of electromagnetism.

Inertia from the Equations of Motion: When M is spatially uniform, the density equation can be rearranged as

$$\frac{P_\mu^2}{2M} = -M\frac{\partial\lambda}{\partial t} \qquad \left(\begin{array}{c}Kinetic\\Energy\\Equation\end{array}\right)$$

The left side is the kinetic energy of the fluid. The right hand side shows overall laminar temporal change of the fluid implied by that motion.

When there are no forces and no charge, just inertia and collisions, the Charge Current Equations and the Kinetic Energy Equation can be joined to reveal basic fluid behavior. To do this take the gradient of the Kinetic Energy Equation and the time derivative of the Current Equations and then add the two equations. The result is:

$$\frac{\partial}{\partial x_\mu}\left(\frac{P_\mu^2}{2M^2}\right) + \frac{\partial}{\partial t}\left(\frac{P_\mu}{M}\right) = 0 \qquad \left(\begin{array}{c}Inertia\\Equations\end{array}\right)$$

When M is a constant, it can be rearranged to give the following more familiar relation:

$$\frac{\partial}{\partial x_\mu}\left(\frac{P_\mu^2}{2M}\right) + \frac{\partial}{\partial t}P_\mu = 0 \qquad \left(\begin{array}{c}Inertia\\Equations\\for\ Constant\ M\end{array}\right)$$

The first term is the work done to the fluid at a location, the second is the force applied to the fluid at that location. Thus, it provides the standard relation between work and force. It says for example, that if the force is zero, then the energy remains constant. The equation is independent of the potential energy terms of the fluid and applies for all ideal fluids.

Expansion of the Universe: The inertia equation can be used to determine the possibility of inertial expansion of the universe. To match observations, we seek, in the equations of motion for the universe, solutions for an expanding universe that provide the following four particular features as observables: (1) spherical shape, (2) expansion, (3) accelerating expansion, and (4) expansion rate equals the local speed of light. The universe will have spherical symmetry if V_μ has only a radial component V_R. Here, hyperspherical coordinates are used: (R, r, θ, φ). R replaces the Cartesian w coordinate; that is, it is the coordinate that is basically perpendicular to observable universes. Also, $R = 0$ references a central origin point from which a hyperverse may expand. The three other coordinates (r, θ, φ) are spherical coordinates for the three observable dimensions.

Light will bend (via refraction) to form spherical observable universes only if the speed of light is proportional to the radius of the observable universe, per Reference 1. Thus, here we seek solutions in which $V_R = RV_T(t)$. Here $V_T(t)$ represents the time dependence of the expansion rate, separate from any spatial dependence; the expansion velocity is the product of its radial dependence and its time dependence. Such solutions were previously found by inspection in Reference 1, so we know that such solutions exist. Here we solve more generally to show how the other features of the universe occur.

First we solve the equations for the possibility of inertial laminar expansion of the universe. Laminar expansion has no turbulence. To determine $V_T(t)$ for inertial expansion, we plug the assumed form for the momentum $P_R = MRV_T(t)$ into the Inertia Equation to obtain a first order non-linear differential equation for the velocity's time dependence $V_T(t)$.

$$RV_T^2(t) + R\frac{\partial}{\partial t}V_T(t) = 0 \qquad \left(\begin{array}{c} Universe \\ Expansion\ Rate \\ Equation \end{array}\right)$$

The solution to this equation is $V_T(t) = 1/t$. Thus, the total expression for the expansion of the hyperfluid is $V_R = R/t$. This solution is the same as one provided in Reference 1, except that here it has been explicitly shown that this is an inertial expansion, and the only inertial expansion that provides spherical observable universes.

For inertial expansion, the fluid density M must be independent of location, i.e. $M = M(t)$. Otherwise forces due to its gradients would occur. Plugging that form for M and the solution $P_R = MR/t$ into the continuity equation creates a first order linear differential equation:

$$(N/t)M(t) + \frac{\partial M(t)}{\partial t} = 0 \qquad \left(\begin{array}{c} Density \\ Rate \\ Equation \end{array}\right)$$

Here, N is the dimensionality of a hyperverse. The solution is $M = M_0/t^N$ for any initial constant value M_0, where for the actual Hyperverse M_0 has some particular value. This is a valid solution so the possibility of laminar expansion exists. Recapping: for inertial expansion, the fluid density of the laminar universe is uniform across all observable universes, and its density is decreasing over time. Also, the speed of light for each observable universe is proportional to its radius so that the same amount of global time is required for light to circle an observable universe independent of its radius. Further, the speed of light is decreasing over time at each location in space.

The radius R_U of any particular observable universe versus time can be computed and used to determine whether its expansion rate is accelerating. Instantaneously, the rate of change of that radius is $dR_U/dt = V_R = R/t$. This rearranges as $dR_U/R = dt/t$, which integrates as

$$R_U = (R_{U_0}/t_0)\,t \qquad \left(\begin{array}{c} Inertial \\ Expansion \\ Equation \end{array}\right)$$

Here R_{U_0} is the radius of some observable universe at some specific initial time t_0. The equation indicates the radius of that observable universe is proportional to time. Thus, for inertial expansion the rate of expansion is constant, not accelerating. Since, the observable universe in the equation is for an arbitrary initial radius, this applies to all observable universes

Observable universes can be truly accelerating if instead of inertial expansion there is an outward force; a force pushing the observable universes outward. Since a uniform fluid density provides a constant rate of expansion, a gradient in the fluid density such that the density decreases away from the center of the expansion can create an outward force that provides an accelerating expansion.

An accelerating universe could, instead of being real, also be an artifact of the use of local time in measurements. (The use of local time in measurements is because existing clocks measure local time). In the expansion equations, time is a global time. The slowing of local time in an observable universe can create an artificial observation of acceleration, as discussed in Reference 1.

Speed of Light is the Speed of the Hyperfluid: Work published in Reference 1 shows that relativistic kinematics in Euclidean hyperspace requires that the flow perpendicular to each observable universe has a speed equal to the speed of light. Analysis of the continuity equation here shows how this occurs. Where the speed of the fluid is uniform, the continuity equation simplifies to

$$\frac{\partial M}{\partial t} + v\frac{\partial M}{\partial x} = 0 \qquad \left(\begin{array}{c} Constant\ Velocity \\ Continuity \\ Equation \end{array}\right)$$

Here, v is the speed of the fluid, x is a coordinate oriented in the direction of the motion of the fluid. Now suppose, as in typical models, waves with a form $exp(ik - \omega t)$. Using such a form for M in the constant velocity continuity equation provides the relation $-i\omega + ivk = 0$, which rearranges as $v = \omega/k$. In standard models the known free-space relation is $c = \omega/k$. Comparing the two results says that the speed of the waves is the speed of the fluid. Further, it implicitly shows that the speed of waves varies in time and space since the speed of the fluid varies in time and space. While this result is for density waves (waves that carry variations in M), the next section shows that the equations for momentum waves likewise match standard models when the speed of the light matches the speed of the fluid.

Traditional Maxwell's Equations: This section shows that equating the speed of light with the expansion rate of the universe (i.e. setting $V_R = c$) explains our view of Maxwell's Equations as time dependent such that $w = ict$. This particular relation between the w coordinate and time has long been thought of as representing a motion along an imaginary direction at the speed of light. The hypervortex model says that it refers to our observable universe's motion along a real w coordinate pushed along by the expansion current.

Consider that, according to the covariant form of the equations, the electric current J_μ creates electromagnetic fields. However, further consider that in a reference frame truly at rest in four dimensions, there would be no electric current – no velocity means $J_\mu = 0$. The reality is that, due to the expansion current, what we call a rest frame is at rest relative to (x, y, z) but it is in motion at the rate V_R along the w coordinate. That is why at rest we still observe an electric current – the current is along w. That is, setting $V_R = c$ creates the current $J_4 = ic\rho$ as needed to transform between covariant and traditional formulations of Maxwell's Equations.

The above explanation is consistent with ideas hypothesized long ago in world-line models. In accord with those models, the hypervortex model shows that the time dependence that we traditionally observe in Maxwell's Equations is due to the motion of the observable universe. The hypervortex model's morphing of world-lines into hypervortices resolves certain details that the world-line models could not.

Free Space Equations of Motion: Plugging the solution for the laminar inertial expansion of the universe into the equation of motions and solving for the Lagrange multiplier λ gives the solution $\lambda = R^2/t$. With that solution the left and right sides of both the density and momentum equations of motion are independently zero. The terms on the right side of the equations are each non-zero yet sum to zero. In contrast, each of the terms on the left side of each of the equations is independently zero. Such is the state of the overall laminar component of the universe. To study small turbulent components added to the laminar universe, we explore the possibility of waves and structures in which the individual terms on the left side of the equations are non-zero while their sum is zero. The equations of interest thus are the free-space equations of motion, since the terms on the right represent the structures in the universe. The free-space equations of motion are:

$$k_M'(M)\left(\frac{\partial M}{\partial x_\mu}\right)^2 + k_M(M)\frac{\partial^2 M}{\partial x_\mu^2} = 0 \qquad \left(\begin{array}{c}\textit{Free Space}\\\textit{Density Equation}\end{array}\right)$$

$$2k_v\frac{\partial}{\partial x_\mu}\left(\frac{\partial P_\mu}{\partial x_v} - \frac{\partial P_v}{\partial x_\mu}\right) = 0 \qquad \left(\begin{array}{c}\textit{Free Space}\\\textit{Momentum Equations}\end{array}\right)$$

To examine these equations first note that any solutions to them, even if they satisfy the equations exactly, only approximately satisfy the real world. To have non-zero values for any of the terms on the left, the momentum and/or the density of the fluid deviate from the laminar solution. In doing so, the terms on the right side of the equation will change. Still, for small waves or other changes to the momentum or density the equations will be satisfied in good approximation.

The Free Space Momentum Equations are essentially Maxwell's equations in free space. They have plane waves as solutions. Moreover, as linear equations they can have waves of arbitrarily small amplitude. Such solutions are used in everyday work to analyze electromagnetic systems and the equations work very well in such cases.

In contrast, the Free Space Density Equation is non-linear. Both terms on the left side of that equation are non-linear. The leftmost term is the product of quadratic dependence on the spatial change of the fluid density and the dependence of k'_M on the density. The next term in the equation is the product of the second derivative of the spatial change of the fluid density and the dependence of k_M on the density. As such, solutions to the density equation are much more difficult mathematically and not of arbitrary size. Solutions to non-linear are very specific and the next subsection exploits this to improve our understanding of gravity.

Gravity Waves: The density equation describes gravity, in particular gravity waves and gravity wells. The existence of gravity waves depends on the ability of the density equation to support them. This depends on the stiffness of the fluid $k_M(M)$. Here we determine the dependence of the stiffness on the density such that it supports gravity waves.

To determine a stiffness that will support gravity waves, we follow the typical assumption that a wave will have the form $exp(ikx)$. If $M = exp(ikx)$ its first derivative is $\partial M/(\partial x_\mu) = ikM$ and it's second is $\partial^2 M/(\partial x_\mu^2) = -k^2M$. Substituting these into the Free Space Density Equation gives $-k'_M(M)(kM)^2 - k_M(M)k^2M = 0$ which reduces to $k'_M(M)M + k_M(M) = 0$. This equation is true if $k_M(M) = 1/M$ and $k'_M(M) = -1/M^2$.

Using the above-derived form for the stiffness $k_M(M)$ the potential energy due to density gradients becomes $\frac{k_G}{2M}\left(\frac{\partial M}{\partial x_\mu}\right)^2$. Here k_G is a constant and the subscript is chosen because of the constant's effect on the strength of gravity. This resolves the unknown in the Density Equation. The Density Equation can now be written as:

$$\frac{k_G}{M}\frac{\partial^2 M}{\partial x_\mu^2} - \frac{k_G}{M^2}\left(\frac{\partial M}{\partial x_\mu}\right)^2 = \left(\frac{\partial \lambda}{\partial t} + \frac{P_\mu^2}{2M^2}\right) \qquad \binom{Resolved}{Density\ Equation}$$

Similarly, the Free Space Density Equation resolves to:

$$M \frac{\partial^2 M}{\partial x_\mu^2} - \left(\frac{\partial M}{\partial x_\mu}\right)^2 = 0 \qquad \begin{pmatrix} Resolved\ Free\ Space \\ Density\ Equation \end{pmatrix}$$

Note that the simple exponential waveform used to resolve $k_M(M)$ allows large waves. It also allows M, the fluid density, to become negative, which is unphysical. Real waves in the real physical world will have amplitudes that are very small compared to the total fluid density.

Gravity Wells: A non-wave solution to the Resolved Free-Space Density Equation exists. The non-wave solution creates a density well, a region of reduced hyperfluid density. The well shape has exactly the form to give the observed behaviors corresponding to the general relativistic refinements of Newtonian gravity. Consequently, this non-wave solution to the density equation is called the gravity well solution.

The gravity well solution for the fluid density equation is $M = M_L exp(-\alpha/r)$. That this is a solution to the Resolved Free-Space Density Equation can be checked by substituting it into the equation. In this non-wave solution, the symbol r represents a 3D radius. Thus, the physical structure corresponding to this solution has infinite length along R, the 4D radius. The symbol M_L is the fluid density for a laminar inertial expansion of a hyperverse. At large values of the 3D radius, the fluid density for this physical structure smoothly transitions to the fluid density of free space.

At the center of the physical well (at $r = 0$) the hyperfluid density becomes zero. That is, these gravity wells have voids at their centers, places where there is no hyperfluid. These gravity well solutions are valid for all values of α. The value of α determines the scale of the solution – larger values of α generate larger scale gravity wells.

Relationships among (a) gravity wells, (b) particle mass, and (c) the observation of 1/r gravity are derived and discussed in upcoming sections. The remainder of this section discusses several features of the gravity well solution in and of itself that are important to that later discussion.

One feature of the gravity well solution, unlike other models, is that the $exp(-\alpha/r)$ form for gravity, is continuous and continuously differentiable everywhere in space, even at $r = 0$. This property is necessary in order to satisfy the principle of relativity at $r = 0$. Exponential gravity satisfies this condition, while General Relativity itself does not.

Another feature of the exponential gravity well solution is that the energy of the gravity well is proportional to the size of the well. Such a relationship is fundamental to the theory of general relativity. However, while General Relativity must postulate this property of matter, here the postulate is replaced with a derivation. That is, this model proves the truth of a property needed by General Relativity, but which General Relativity cannot itself prove.

To derive the relationship between the energy of the well and the size of the well, the first step is to compute the potential energy of the physical structure described by the mathematical solution for the gravity well. The potential energy per unit length of a vortex due to a gravity well within that vortex is given by the integral of the potential energy term $\frac{k_G}{2M}\left(\frac{\partial M}{\partial x_\mu}\right)^2$ for the gravity well solution $M = M_L \exp\left(-\frac{\alpha}{r}\right)$. To simplify the integration, the standard mathematical relation is used that: for any function Y the relation $\frac{\partial ln(Y)}{\partial x_\mu} = \frac{\partial(Y)}{\partial x_\mu}/Y$ applies. Using this relation, the potential energy term becomes $\frac{k_G M}{2}\left(\frac{\partial ln(M)}{\partial x_\mu}\right)^2 = \frac{k_G}{2}exp\left(-\frac{\alpha}{r}\right)\left(\frac{\alpha}{r^2}\right)^2$. Integrating this over r, for r from zero to infinity, in spherical coordinates, gives $\left[\frac{k_G}{2}\alpha exp\left(-\frac{\alpha}{r}\right)\right]_0^\infty = \frac{k_G}{2}\alpha$.

The above equation shows that the gravitational energy per unit length of a vortex that contains a gravity well is proportional to α. Since the strength of the gravity well is also proportional to α hence the potential energy and the size of the gravity well are proportional to each other. Further, the equipartition theorem says that energy is distributed to the various quadratic parts of the Lagrangian in proportion. Therefore the total energy of a vortex in the fluid is proportional to the size of the gravity well, for vortices that include gravity wells as part of their structure.

A third feature of the exponential gravity well is that the gravitational potential created by a set of gravity wells is the sum of gravitational potentials of the individual gravity wells. General Relativity and other gravity models cannot prove this feature, and thus must postulate it. For exponential gravity wells, the proof follows.

To start the proof, consider two gravity wells nearby each other. Their combined impact on the fluid density is given by the equation $M = M_0 \, exp\left(-\frac{\alpha_1}{|r-r_1|}\right) exp\left(-\frac{\alpha_2}{|r-r_2|}\right)$. Here, the subscripts identify the two gravity wells. Amazingly, this complex mathematical form is an exact solution to the Resolved Free-space Density Equation. When there are n gravity wells, the solution to the Resolved Free Space Density Equation is $M = M_0 \prod_{i=1}^{n} exp\left(-\frac{\alpha_i}{|r-r_i|}\right)$. Here α_i and r_i are the size and location of the ith gravity well in the set. The above form can be rearranged to form $M = M_0 \, exp(-\sum_{i=1}^{n} \alpha_i/|r - r_i|)$, which completes the proof.

Equivalence of Inertial Mass and Gravitational Mass:
The relation between inertial mass and energy was discovered when Einstein wrote $E = m_0 c^2$. Here, m_0 is the inertial mass of a particle when it is at rest. The hypervortex model shows that, physically, the energy, E, is the sum of the kinetic and potential energy of the hyperfluid in the portion of the particle's hypervortex that intersects our observable universe. Thus, the inertial rest mass m_0 of a particle is proportional to the energy per unit length of the particle's hypervortex.

The relation between the size, α, of a gravitational well and the energy of the corresponding hypervortex was derived in the previous section. It showed that α is proportional to the energy per unit length of the hypervortex. Combining this proportional relationship with the one above for inertial mass, it is clear that inertial mass is proportional to the size of the gravity well. That is m_0 is proportional to α. Further, comparing the exponential form for gravity to the standard form for gravity, it is clear that $\alpha = gm_g$, where g is the gravitational constant and m_g is the gravitational mass of observed particle. Hence, $m_0 = m_g$. That is, inertial mass equals gravitational mass.

Observing $1/r$ Gravity: Exponential gravity is consistent with our observation of $1/r$ gravity. Indeed, it was shown decades ago to reproduce the observed results of General Relativity. There are several possible interpretations in which the exponential and $1/r$ forms for gravity may provide the observed results. In one interpretation both the exponential and $1/r$ forms provide identical results for all observables. In another interpretation the two forms provide nearly identical results except where r is very small.

The latter interpretation is common in previously published results regarding exponential gravity. Despite the very different mathematical formulas for exponential and $1/r$ forms for the gravitational potential, the gravitational forces that are computed as the derivative of each formula are nearly identical at locations where we are typically able to measure gravity.

The hypervortex model adds the interpretation in which the exponential and $1/r$ forms provide identical observable results for all values of r. In particular, the hypervortex model provides the possibility that the exponential gravity formula in global Euclidean coordinates, when transformed into the local curved-space-time coordinates of General Relativity, transforms exactly to a $1/r$ formula for gravity. That is, if from a global perspective with its constant ruler (which we do not actually possess or have access to) gravity is exponentially related to the distance from the center of a gravity well, then in measurements performed with a local ruler (which is the type that we do have) any observer measures gravity locally as being $1/r$.

The derivation showing that exponential gravity in global coordinates is the same as $1/r$ gravity in local coordinates is not included here; a description of how to do the derivation is provided. Specifically, wherever an observer is in a gravity well, the ruler is appropriately shrunk relative to a ruler away from the gravity well. The shrinkage is such that the observer measures gravity as $1/r$. To derive this, first realize that the ruler is typically a beam of light. Also, curvature in General Relativity corresponds, in Euclidean coordinates, to bending of light. In particular, the variation in M in the gravity well causes variations in the speed of light which causes refraction that we observe as gravity. This is used to determine the radial dependence of the size of the local ruler at various distances from the center of the gravity well. Once the radial dependence of the ruler is known, the shrinkage of the ruler can be corrected for to get the correct global form for gravity. That form is the exponential form provided in the previous section.

The exponential form for gravity has been used to get correct results for observable non-Newtonian gravitational effects including the bending of light around a star and the advance of the perihelion of Mercury. Those derivations are not presented here as such derivation would duplicate results published elsewhere. The point here is that the hypervortex model provides such solutions starting from a unified Lagrangian density that derives fundamental properties of matter that other models must assume.

$1/r$ local gravity has problems satisfying the principle of relativity at $r = 0$ and at the Schwarzschild radius of black holes. At $r = 0$ the $1/r$ form has a mathematical singularity, gravitational energy becomes mathematically infinite and the principle of relativity is not satisfied. The Schwarzschild radius has another unique problem in that light cannot travel in all directions there, which certainly violates the principle of relativity. In contrast, for exponential gravity in the Euclidean frame, relativity applies everywhere. At the Schwarzschild radius, the speed of light does become exceedingly slow, yet an observer at the Schwarzschild radius will see light travel at the same slow speed in all directions. That does mean that the hypervortex model says that light can escape black holes, a concept with which various researchers using variant curved coordinates concur.

The Speed of Sound in the Hyperfluid in Free Space:

Whereas light waves are transverse waves in the hyperfluid, sound waves in the hyperfluid would be associated with longitudinal waves. Maxwell thought of light as being the equivalent of sound, but we have seen that it is not. This section examines the possibility of waves in the hyperfluid, but only in free space where the fluid flow is laminar, i.e., away from any vortices.

The speed of sound in any medium depends on its compressibility, β, and its density, ρ. The speed of sound is well studied for fluids because of the importance and prevalence of air and water in our environment. Once we know a fluid's compressibility and density, its speed of sound is computable by the general equation $(\beta\rho)^{-1/2}$. The equation says that the speed of sound is inversely proportional to a fluid's compressibility and its density. That is, higher density and compressibility mean a lower speed of sound. Maxwell, in his fluid model obtained an equation for the speed of light as $(\epsilon_0\mu_0)^{-1/2}$. His solution has the same overall form as the equation for the speed of sound and thus he thought, that light was analogous to sound.

Compressibility of a fluid depends on its structural properties. Computation of the hyperfluid's compressibility uses the definition of compressibility as the inverse of the change in pressure for a change in density. For the hyperfluid in free space, pressure is proportional to $\frac{k_G}{M}\left(\frac{\partial M}{\partial x_\mu}\right)$. If a quantity of hyperfluid is compressed into half the volume, both the density M and the gradient $\left(\frac{\partial M}{\partial x_\mu}\right)$ double, which means that the pressure is unchanged. That is compressing hyperfluid does not change its pressure.

To understand the implications of the above result, note that in air, doubling the density basically doubles the pressure. Further, in water, which is less compressible than air, the density of water cannot be doubled. Thus, hyperfluid is more compressible than air and water. In fact, the equation says that hyperfluid in free space is infinitely compressible. Since, as discussed earlier, the speed of sound is inversely related to compressibility, hence infinite compressibility means zero speed for sound waves in free space in the hyperfluid. Thus, sound waves, at least as we understand them from studying fluids on Earth, do not exist in free space in the hyperfluid, which would explain our lack observation of such waves. Full solutions to the full hyperfluid equations may expose other wave types in free space.

Entanglement Waves in the Hyperfluid: While the speed of sound in free space in the hyperfluid is zero because of infinite compressibility, the speed of sound in the hyperfluid will be high in places where structure in the hyperfluid provides substantial stiffness; stiffness is the inverse of compressibility. Substantial data supports the idea that communication among entangled particles is enabled by a high speed wave along the length of vortices enabled by stiffness of vortices regarding compression, tension, and torsional forces.

The data that provides evidence of hypervortex stiffness is the complete body of data showing that the measured mass of stable particles is the same for all particles of a given type, within measurement accuracy. The constancy of the measured mass of such particles indicates that the mass per unit length of such particles is very difficult to vary even by a small amount. Such stiffness is much higher than the stiffness of hyperfluid in free space. At the same time, as was shown in the section regarding gravity, the density of hyperfluid is lower inside vortices than in free space. Thus, per the equation given in the previous section for the speed of sound, the speed of waves that stretch, compress, or even apply torsion of the hypervortex will have very high speed. Such waves are the types of waves that would be created in experiments regarding entangled particles. Thus, such waves are likely the mechanism of "Spooky Action", i.e. the mechanism for communication among entangled particles. To quantify the velocity of such waves requires solving for the full detailed the structure of elementary particles, which is discussed in the following section.

Elementary Particle Structure: Elementary particles are the observable section of localized structures (typically vortex structures) in the hyperfluid that satisfy all of the equations of motion, absent the presence of other structures, but including the expansion of the universe. The solutions must satisfy the full equations, not the free space approximations. When the particle structure is correct, observable properties of particles will be computable from the structure, and the values obtained by such computation will match the observed values. Values for electric charge, as computed from the particle structure via the Charge-Density Equation, will match the observed values. Also, the particle mass computed as the total energy of the vortex per unit length will match observed values.

Computation of detailed exact vortex structure for an elementary particle in the hyperfluid might be possible analytically in the absence of the expansion current; indeed such solutions have been explored with some success. However, including the expansion current increases complexity of the task.. It may be less complex than computing the structure of a tornado or a hurricane; however, it is still a daunting task. Detailed tornado and hurricane modeling requires super-computers and, despite decades of effort by many researchers, is an incomplete research task. Also, the need to model four dimensions for the hyperfluid increases the required computational resources by at least several orders of magnitude. In that one regard, the mathematics for the hyperfluid is much more intense than that of weather modeling. Thus, this section presents only a path to performing the computations of vortex structure, not final answers.

The path to solving for particle structure leverages certain properties of the equations, as revealed in prior sub-sections of the chapter, plus certain properties known from the real world. First, from the equations, the density and momentum equations each separately support a continuum of solutions. Second, from both the equations and the real world, particles with mass include gravity well solutions in their overall solution to the coupled equations. Third, from the equations, massless particles, while having no gravity well solution, still have a gravity wave component in their solution. Fourth, from the real world, elementary particles come in discrete sizes – discrete charge, mass, spin, and other properties. These properties, together with general experience regarding the properties of solutions to non-linear mathematics, suggest at least one path to refine and solve the hyperfluid equations of motion for elementary particles whose properties match experimental results.

To explore the suggested path to exact solutions, consider first, as a simplified case, a small gravity (density) wave with no corresponding momentum wave. Such a density wave may have the form $M = M_L exp(\psi)$, where M_L is the density profile for the laminar expanding universe. For purposes of this simplified case, further assume ψ has the form $\psi = 1 + \alpha\, exp(ikx - \omega t)$. When substituting this form for M into the terms on the left side of the Resolved Density Equation, several terms cancel; the residual term is $k_G \frac{\partial^2 \psi}{\partial x_\mu^2}$. This residual on the left side of the equation will act as a forcing function to drive the right side of the equation to deviate from zero.

By the assumption above that the small density wave has no corresponding momentum wave hence the residual on the left must be compensated by a change to λ in the equation. To examine that response define $\lambda = \lambda_L + \Delta\lambda$, where λ_L is the laminar expansion of a hyperverse. Then the Resolved Density Equation becomes approximately $k_G \frac{\partial^2 \psi}{\partial x_\mu^2} = \frac{\partial \Delta\lambda}{\partial t}$. The term involving P_μ in the equation drops out because its form for the laminar expanding universe matches and cancels the term for λ_L . This form of the equation is not yet useful because there are two variables ψ and $\Delta\lambda$, but only one equation, which is insufficient to provide a solution for either ψ or $\Delta\lambda$.

To create a second equation, basic properties of physics are used that are valid for small perturbations, and often also for large ones. First, as a residual, the term on the left becomes a forcing function for the term on the right. Forcing functions are a well understood aspect of physics in general, and they imply that the term on the right of the equation will be 90 degrees out of phase with the forcing function. Second, the response function generally follows the forcing function in shape. Third, the normalizations of ψ and λ are not necessarily matched. These three differences in combination give an approximate forcing-function relation $\psi = -iK\Delta\lambda$. Here $-K$ represents the unknown and adjustable ratio of amplitudes of ψ and λ. The symbol "i" is the imaginary number that is used as a standard technique to put $\Delta\lambda$ $90°$ out of phase with ψ in a mathematical representation that allows the use of complex numbers to solve the equation. Using this additional relation to remove $\Delta\lambda$ from the previous equation gives finally,

$$\frac{\partial^2 \psi}{\partial x_\mu^2} = \frac{-iK}{k_G}\frac{\partial \psi}{\partial t} \qquad \binom{Unnormalized\ Free\ Space}{Schr\ddot{o}dinger's\ Equation}$$

This equation is labeled the Un-normalized Free Space Schrödinger's Equation because it is formally similar to that equation. That similarity is discussed further in a later section. Here, an important point is that any solution to this still adds a change $\Delta\lambda$ to the momentum equation. Thus, there will be a momentum effect. Even in the simplified case, the solution is in conflict with the assumptions of this case. Thus, by contradiction, massless particles include both a gravity (density) wave and an electromagnetic (momentum) wave.

Now consider a more general case that may include changes to the momentum and which may include a combination of a gravity wave and gravity well as a solution to the Resolved Density Equation. The gravity well solution does not change the residual on the left side of the Resolved Density Equation; it does however cause large changes to the fluid density, which, in turn, creates large changes to the P_μ/M terms on the right side of both the density and momentum equations. These changes couple the two equations. Using the presence of P_μ/M on the right side of both equations to remove it, the two equations can be merged as:

$$\frac{\partial^2 \psi}{\partial x_\mu^2} = \left(\frac{\partial\lambda}{\partial t} + \frac{1}{2}\left[\frac{\partial\lambda}{\partial x_\mu} - 2k_V\frac{\partial}{\partial x_\mu}\left(\frac{\partial P_\mu}{\partial x_\nu} - \frac{\partial P_\nu}{\partial x_\mu}\right)\right]^2\right) \qquad \binom{Nonlinear}{\substack{Particle \\ Equation}}$$

To solve for vortex structure, this equation must be solved together with the continuity equation. Since there are three unknowns and two equations, it may seem that an infinite continuum of solutions must exist, even for the smallest particles. However, these are non-linear equations; there will be discrete solutions. Moreover low energy solutions will be relatively farther apart than higher energy solutions. The equation is quite general; the chosen form for the density $M = M_L \, exp(\psi)$ for any real ψ does not appear to restrict the solution set in any way. To solve it, various relations among the three variables can be assumed to create a third equation. All exact solutions found for all such third equations are real. Different types of particles may satisfy different forms for the assumed third equation.

The relation $\psi = -iK\Delta\lambda$ is suggested as a third equation both because of the basic validity of the physics used to generate it and because its earlier use generated the Un-normalized Free Space Schrodinger's Equation. It can be used together with $\lambda = \lambda_L + \Delta\lambda$ in the non-linear particle equation to completely remove λ from the calculation. Also, an additional similar assumption $P_\mu = P_\mu^L + \Delta P_\mu$ looks promising for use in combination with the above assumptions for λ and ψ . Using it with the assumed form for density $M = M_L exp(\psi)$ in the continuity equation, the equations reduce to two equations involving only ψ and ΔP_μ . If there are solutions, those solutions will be exact; the use of these relations do not themselves make the solutions approximate. Rather use of the relations reduces the searched solution space. If there are no particle solutions using the forcing function relation, perturbation approaches can be used to seek exact solutions for particles whose relation between ψ and $\Delta\lambda$ approximates the forcing function relation.

If current quark models of elementary particles are correct, then leptons, mesons and baryons will each internally contain one, two, or three vortices, respectively, and a corresponding number of gravity wells. The assumed form for ψ, thus, will be a combination of the appropriate number of separate gravity wells (and gravity waves perhaps). Alternatively, since the Milky Way has one vortex (with a couple of small ones orbiting it) despite its astronomic scale, it may be that all elementary particles are each one vortex. In such a case, the symmetries that the current models expose may map to some other aspect of ψ, or to some aspect of P_μ.

Some standard models add myriad dimensions of miniscule physical extent to explain the set of elementary particles. No such need has been identified yet in the hypervortex model; if such myriad dimensions are necessary, that need is most likely to appear in this aspect of physics. If such need does occur, that does not impact the validity of the hypervortex model. Rather it refines our knowledge of its dimensionality. Similarly, if the calculation of the structure of elementary particles exposes a need to refine the properties of the hyperfluid that too is not a flaw in the model, but rather a reason to bother with calculating detailed particle structure. Any such refinements require a change to the Lagrangian, and then re-derivation of the equations of motion. That may lead to new terms in the Non-Linear Particle Equation, which in turn will adjust the particle structure. Iteration may occur until the particle charges and masses match observations. Such a process is similar to the means used to create the existing standard models for elementary particles.

Strong Forces: Standard models have not, despite many years of research by many scientists, identified an exact quantitative form for the strength of the strong forces. Thus, do not expect such achievement here. Rather, this section presents a path to computing the strong forces for various particles and circumstances.

In the hypervortex model, the strong force is the difference between the actual force and the force predicted by application of the gravitation and electromagnetic interaction equations. Based on observations, these differences occur only for certain particles and only near the vortex axis of those particles. The suggested process to compute strong forces uses the Hamiltonian representation of the physics, rather than the Lagrangian approach. This is due to both the fact that it tends to follow the Schrödinger approach used to compute electromagnetic coupling in the near-field, and because the Hamiltonian comes with equations for computing forces.

The Hamiltonian reuses the physics used in the Lagrangian. No new physics need be added; only the mathematics changes. Whereas the Lagrangian examines the difference between the kinetic and potential energy and the sloshing of energy between these two forms, the Hamiltonian approach uses the sum of the kinetic and potential energy and the fact that this sum is conserved. Given a Hamiltonian H of a system, the equations it provides are:

$$\frac{\partial H}{\partial x_\mu} = \frac{dP_\mu}{dt} \quad and \quad \frac{\partial H}{\partial P_\mu} = \frac{dx_\mu}{dt} \qquad \begin{pmatrix} Hamiltonian \\ Equations \\ of\ Motion \end{pmatrix}$$

The first of the two above equations directly relates the force to the Hamiltonian. The second equation is not used in the process suggested here. The first equation shows that, given the Hamiltonian for a particular configuration of particles, the derivative of that Hamiltonian with respect to the appropriate change in the location of a particle is the force on that particle. That is, dP_μ/dt is the force.

To use the Hamiltonian Equations of Motion to solve for strong forces create a combined Hamiltonian for the vortices of interest. The first step is to compute the structure for each of the particles in free space. In the hypervortex model, this amounts to solving the Non-linear Particle Equation for the appropriate vortices individually, including the laminar expanding universe. The solutions might be numerical, as from a computer, or analytic in the form of an exact equation. Given the individual solutions, the combined Hamiltonian can be created. Whereas the solutions for the individual vortices will be centered at the origin of the coordinate system, the combined solution must move one or more of the individual solutions away from the origin.

The individual solutions may be combined as sums or products, depending on mathematical perturbation methods chosen. The methods are formally similar to perturbation methods that have been successfully applied with Schrödinger's equation except that the non-linear aspects must be considered. Also, it was shown earlier in the chapter that, for gravity well solutions, composing a solution for multiple vortices as a product of exponentials works well, even including the universe expansion.

In the format shown for a product of multiple gravity wells, the ith gravity well was located at a location r_i. To then compute the force on the ith vortex, the derivative of the Hamiltonian is taken with respect to r_i. The derivative may be taken analytically or numerically (via computer). The strong force then is the difference between the force computed via this method and the force computed via the equations for gravity and for electromagnetic force.

Data indicates that only hadrons have a strong force. The hypervortex model can be checked to see that it matches the data. To do so the Hamiltonian can be constructed for non-hadronic vortices (electrons for example) and the derivative taken. If there is no strong force for such particles, then the force computed from the Hamiltonian Equations of Motion should equal the sum of the electromagnetic and gravitational forces.

The hypervortex model can be used to investigate the possibility of symmetry breaking in hadrons. For example, the Hamiltonian can be constructed separately for particles and for the corresponding antiparticles. If there is no symmetry breaking for a given configuration, the results will be the same for the particle and anti-particle. Investigation of possible symmetry breaking in hadrons can be expanded by examining the symmetry of the Hamiltonian with a galactic-scale vortex included, i.e. a "super-vortex". This could, for example, examine the possible effects of our galaxy on the symmetry properties of hadrons. Such analysis can be further generalized by including various types of super-vortices. Some super-vortices might create symmetry breaking while others do not.

The strong force on a particle may depend on its orientation relative to the other particles within its proximity. When such orientation is not known, as is typical in real-world situations, the strong force can be estimated by using the Hamiltonian approach to compute the force for all orientations and averaging.

Weak Forces: The weak force changes the structure of vortices. This can be studied in the hypervortex model via computational methods that simulate the time evolution of vortices using the Lagrangian or the Hamiltonian. The computation can be initialized using the time independent solution for a vortex, and then evolved over time. Computer simulations are a good method for this – similar to modeling the motion of a hurricane. Structural changes may be induced by mathematically applying various impulses or collisions. Photons and vortices can be added to the initial configuration such that they will collide with the vortex of interest. This is not dissimilar from methods for modeling a hurricane colliding with a weather front. It is known that weak forces have symmetry breaking. This is similar to the fact that hurricanes rotate counter clockwise in the Northern Hemisphere because the Earth rotates counter clockwise in the Northern Hemisphere. Similarly, the rotation of the equivalent super-vortex must be included to model weak interactions of vortices.

The mathematical appearance of the Lagrangian for weak interactions, when including the rotating super-vortex, may have similarities to the mathematics of the weak interaction for standard models. For example there will be an angle between the rotation of the super-vortex and the elementary-particle vortex that determines the degree of symmetry breaking. The difference is that the hypervortex model provides a root cause for the occurrence of the symmetry breaking. Also, in the hypervortex model the universe has symmetry; the symmetry is broken for particles, just as for tornadoes, by the local environment which both creates the vortices and then affects their stability and behavior.

Quantum Mechanics: The mathematical formalism of quantum mechanics fits unchanged into the hypervortex model. This section discusses two quantitative aspects – the Heisenberg Uncertainty Principle, and the Schrödinger Equation. It shows that the Heisenberg Uncertainty Principle summarizes aspects of how actual vortex paths deviate from classically computed particle paths. It shows that the Schrödinger Equation is likely a linear approximation of the full non-linear hyperfluid equations that works when such approximation is appropriate.

In the hypervortex model, the Heisenberg Uncertainty principle indicates that vortices have surplus vortex length. Like a string connecting two points, but with extra length, the surplus length allows the vortex to choose among a wide variety of paths between two connected points. In contrast, classical physics is equivalent to a string without surplus length. Once we interpret the Heisenberg Uncertainty Equation as indicating that vortices have surplus length, its equation $\Delta x_\mu \Delta p_\mu \geq \hbar/2$ becomes a specification of how far a hypervortex path can deviate from the classical path, enabled by that surplus length. Note that "p", different from the fluid momentum "P", denotes the observed momentum of an observed particle. Together with the Schrödinger Equation and other factors, the Heisenberg Uncertainty Equation enables computing the properties of the paths that vortices may take.

The hypervortex model provides the reason that the actual vortex path deviates from the classical path. The classical path is computed using the physics of gravity and electromagnetic as separate forces. However, it was shown in several sub-sections above that, in actual solutions for particles, the two are coupled. This creates, as described in an earlier sub-section, a forcing function that drives the energy to slosh between fluid density and fluid momentum. The effects of the forcing function are not included in the computation of the classical path. Such a forcing function creates a periodic oscillation, in general, and there is nothing apparent about the fluid to create an exception in this case. Thus, the coupling creates an oscillation in the vortex. The Heisenberg Uncertainty Relation summarizes the size of the oscillation and the Schrödinger Equation computes the frequency and wavelength of the oscillation.

The shape of each individual vortex path depends on the type of path taken by the actual vortex. Consider vortices whose actual path forms a helical orbit around the classically computed path. In such case the center line of the helix is the classical path. The radius of the helix is Heisenberg's Δx. For non-relativistic motion, the corresponding deviation in the momentum is related to the angle between the helix center line and the helix via the equation: $\Delta p = sin(\Delta \theta)mc$. The symbol m is the particle mass and c is the speed of light. The relation between the angle and the momentum comes from the basic kinematics of vortices in hyper-space, as discussed in the beginning of this chapter. The Heisenberg Uncertainty Relation, in combination with the relation between frequency and wavelength for the vortex as provided by the Schrödinger's Equation, provide sufficient information to determine relations among the quantities: Δx, Δp, the wavelength of the helix, and the frequency of its oscillation. The wavelength is not in general the same as the Compton Wavelength or the de Broglie Wavelength of other historical interest, though they are related in some cases. The detailed derivation has been written up for separate publication. Its details are interesting but not fundamental to the hypervortex model. That is, they occur whether the root cause is a hyperfluid or something else. The properties of this helical orbit suggest that it directly relates to the property of the particle called "spin". In contrast, the overall rotation of the vortex is the electric charge rather than the spin.

The available information in any real world situation is, in general, insufficient to compute the phase or the eccentricity of the orbit of any particular vortex around the particle path predicted by classical physics. The helix may have a left or right rotation. Except when some force makes one state preferred, the vortex may be in a mix of states. Thus, the actual path need not be circular. It could be ovoid or even linear. The possibilities are describable as a linear combination of left and right coiled helices. Because we can only observe a small piece of the vortex intersecting the observable universe, we cannot know the exact path that any vortex is taking, nor can we know what state it is in. Or at least we have not found any way to enable such observation to date. The hypervortex model indicates that, no matter how well we might measure the location and momentum of a particle at one time, we still do not know where the particle will be later because the measurement does not identify the path of the center of the helix, the phase of the vortex in its orbit or the eccentricity of the orbit. Thus, no matter how much our measurement may avoid affecting the particle, after a measurement, our knowledge of the location and momentum of the vortex is at best statistical.

The Schrödinger equation provides the detailed mathematics for computing quantum scale motion of particles. Interpretation of the Schrödinger equation and its relation to the hyperfluid equations of motion is clarified by the hyperfluid interpretation of the Heisenberg Uncertainty Principle. In particular, it appears now that Schrödinger's contribution amounts to mathematical machinery that solves the hyperfluid equations of motion in a useful subset of circumstances. The previous section introduced the Hamiltonian approach for exact non-linear calculations of vortex interactions. Schrödinger appears to have adapted the same approach to examine vortex behavior statistically for small perturbations in non-relativistic circumstances.

Schrodinger built his approach phenomenologically. Also, he built it with a deliberate goal of creating a linear equation. Decades of data and analysis shows that it works predictably for a range of conditions and not outside those conditions. With the hypervortex model providing new insight, a formal derivation of the Schrodinger Equation as a linear approximation of the full hyperfluid equations appears possible and valuable. Such derivation would provide improved understanding of its range of valid use. It might also reveal additional details of hyperfluid behavior.

It was originally intended that the book map out a path to achieving a formal derivation of Schrodinger's Equation. Then a check of prior work in Reference 5, confirmed that Hamilton and Jacobi (another key mathematical physicist) had already worked out key necessary mathematics towards such formalism. They showed that solving the Hamiltonian Equation for a system is equivalent to finding the path of a string between two points. They also showed that a string with excess length satisfies the un-normalized Schrödinger Equation provided above, although they did not call it that. Thus, all that needs to be said here about a path to a formal proof is that it follows the path of those efforts of Hamilton and Jacobi.

9. Life, Free Will, and Determinism

The discussion up to this point has hopefully convinced the reader that the hypervortex model provides a framework for unification that addresses, or potentially addresses, the limitations of standard models. If so, then the astute reader may at the same time have also noticed one overarching concern regarding the hypervortex model itself. In particular, the hypervortex model raises the following questions:

"How can each observable universe, so intertwined with a near infinite stack of correlated observable universes, present to us cause and effect, and free will or even the appearance of the opportunity for free will?"

"How can we, whose bodies are comprised of hypervortices that stretch through this near infinite stack of correlated observable universes, observe only one such universe, and can we, or any being, change the paradigm to observe and even affect multiple observable universes?"

These are significant questions. They are challenges to the hypervortex model. The possibility of a path to successful answers to these questions is important to the viability of the model. If the model cannot provide such a path, then the model is suspect. At the same time, if it can, it opens new scope for science. This chapter explores the questions. It presents the work to date to check that the model does provide at least reasonableness arguments to successful answers to these questions.

The Possibility of Free Will: The hypervortex model tends to confirm the idea that Quantum Mechanics is the foundation of free will. What the standard models call "space-time" splits into two separate aspects in the hyperfluid. One part is the extra spatial dimension along which infinite numbers of observable universes correlate with each other as past, present and future. The other part is the time coordinate that enables the configuration of the Hyperverse to change over time. The passage of time that we observe is a combination of the two. The extra spatial dimension provides the deterministic part of our experience. The time coordinate provides interactivity with the hyperfluid that allows us to change the configuration of the Hyperverse, thus changing the deterministic future laid out for us. Thus, it is the time coordinate that provides free will.

Consider an analogy to frames in a movie. One frame of the movie is a view of a selected person in our 3D observable universe. For some large number of adjacent 3D observable universes, a snapshot of the same person in each 3D universe provides a frame for the movie. Since the observable 3D universes are spatially related as past, present and future, a full set of snapshots, shown in sequence, presents a movie of the deterministic part of a piece of that person's life – past, present and future. The validity of such a movie is limited by two things: (1) relativistic motion, and (2) free will. Absent free will, as our observable universe flows along with the expansion current, we observe the sequence of frames of the movie one frame at a time.

Relativistic motion is motion at speeds approaching the speed of light. The impact of relativistic motion on the frames of the movie is a detail of great importance for other discussions, but not as regards free will. In the analogy, relativistic motion is approximately akin to a blur that occurs when filming high speed objects. It limits our ability to view the deterministic portion of the past and future in the movie.

Free will lets us change the content of the frames and thus control the movie. More specifically, it is analogous to the ability to edit a frame in the movie as it plays and such that the edit propagates changes to all subsequent frames of the movie. Since the time coordinate is necessary to enable free will, the aspect of the hyperfluid behavior that enables free will can be identified by finding where the time coordinate explicitly appears in the hyperfluid's equations of motion. Such analysis shows that quantum mechanics, and only quantum mechanics, explicitly invokes the time coordinate and therefore it is the enabler of the possibility of free will. In contrast, the free-space equations for electromagnetism do not explicitly contain time, only the extra spatial coordinate masquerading as time. Solutions to that equation create stationary solutions in the Hyperverse. We observe those stationary solutions frame by frame and think they involve time. Similarly, the free-space equation for gravity makes no explicit mention of time. Solutions to it in isolation create stationary solutions in the Hyperverse that we observe frame by frame. Only quantum mechanics, as a manifestation of the effect of the coupling among the equations, involves real temporal evolution of the configuration of hyperspace. It allows us to initiate a change in the configuration of our observable universe and, over time, propagate that change throughout the Hyperverse.

Note however, that quantum mechanics only creates the possibility of free will. It is a necessary factor for the realization of free will. It alone is not sufficient to realize free will. A rock does not have free will. In addition to requiring a real time coordinate, and quantum mechanics, free will also requires: (1) an ability to observe the flow of past and present; (2) an ability to consider the expected future that will occur if no action is taken; (3) an ability to devise and execute a preferred alternate action; and, (4) a set of mechanisms by which the alternate action is sufficiently powered to impact the whole of the Hyperverse.

Observing the Universe: Our observation of the Hyperverse as a sequence of 3D observable universes seen one at a time occurs if the nature of consciousness is fundamentally photonic, i.e., electromagnetic, as has been previously discussed. It is photons (and other massless particles) that are trapped and separated into 3D observable universes. Thus, the essence or our existence is not corporeal. Our corporeal body, being composed of vortices, has substantial length along the extra spatial dimension. We each observe ourselves as one 3D corporeal body; however, for each of us there exists a near infinite set of 3D corporeal bodies, one in each of the observable 3D universes that span our lives. Those bodies, like other objects, are approximately configured at any moment to trace out the deterministic aspects of our past, present and future. The corporeal aspect of our existence serves many good functions, but it is not the essence of consciousness. It cannot be, according to the hypervortex model, because if it were, we would not observe 3D observable universes one at a time.

The above realization led to a search for a physical structure within life forms that could sustain, trap and interact with a fundamentally electromagnetic consciousness. That structure appears to be DNA. This was hypothesized and explored with experiments and with calculations. The key component in DNA that enables its function is the stack of base pairs at the center of the double helix, not the helix itself. The double helix is a mechanical frame on which to hang the base pairs. The base stack in DNA is normally a weak electrical conductor, a stack of quasi-one-dimensional graphite-like planar surfaces that conduct electricity much as it is conducted in graphite. However, in appropriate circumstances, a synchronized combination of electrical and mechanical vibration modes will cause temporary 90 degree rotation of groups of base pairs. In the rotated state, a string of base pairs provides electrical conductivity better than any metal. Some measurements claim to observe superconductivity in DNA. These high conductivity regions propagate along the DNA molecule as waves. The threshold energy in the DNA to initiate the waves has been measured to agree with computed estimates. Also, measurements and computations agree on a speed of these waves of about 3 kilometers per second, which is about $1/100,000^{th}$ of the typical speed of light. Their slow speed is important – it slows the electromagnetic waves so that they can be contained inside an organic cell, inside a DNA molecule, as an electromagnetic consciousness.

The hypervortex model contributed to the efforts to understand the electromagnetic properties of DNA. Published literature includes many investigations into DNA's electromagnetic properties; however most are empirical without the hypervortex model to guide them. Lacking such guidance, results regarding DNA's electromagnetic properties have been controversial and difficult to reproduce. For example, early DNA studies successfully measured the correct speed of electromagnetic waves inside DNA, but then later, researchers could not reproduce those results. Also, the DNA theories that existed at the time could not explain the results. Later, using the hypervortex model as a guide, new experiments were able to reproduce the original results. The new results also explained the prior difficulties including the prior failure to reproduce early results. In particular, the new experiments showed the ability to make DNA interact with electromagnetic fields, or not, depending on details of the configuration of the experiment. New computations leveraging the knowledge provided by the hypervortex model refined the understanding by showing that a particular type of vibration mode in DNA was central to the ability to vary the electromagnetic properties of DNA. The earlier computations did not find the right vibration mode and thus their computational results did not match the experimental results. The hypervortex model guided the effort to find the additional vibration mode – the mode whose behavior is such that life observes one 3D observable universe at a time.

The Nature of Thought: Thought processes, like the process of observation, are localized into 3D observable universes. Otherwise our observation of event sequences and causality would be different than it is. Thus, DNA is the likely candidate as the central processing unit of organic life. Such claim may sound initially contradictory to neurological models of the brain; however the two views are views of different levels of thought. DNA provides processing power to each cell in an organism, neurons allows the processing power of the cells to be combined.

There is substantial analogy between computer architecture and the relation between DNA and neurons. In computer technology, one or more processors comprise a physically small part of each computer and the remainder of the computer provides power, protection, and interfacing to peripherals, sensors and other computers. Large scale computer centers contain thousands of processors linked in networks. The networks and their organization determine the real power of the large scale computer centers, but at the heart of each computer is still its little processors. DNA strands are analogous to the individual computer processors, the neurons are analogous to the network and other information relay methods, and the brain is like the large scale computer center.

The DNA molecule provides a conduit that does nothing until electromagnetic fields run through it. As with electronic data processors, the physical device is only a part of the processor. It sits there, dead, until power is applied that creates the time varying electromagnetic fields to be manipulated by their interaction with each other and with the physical device. Thus, inside an active DNA molecule resides a trapped laser-like beam – an electromagnetic wave. Each DNA molecule sustains one such coherent photonic wave that spans the length of the molecule. Depending on the type of DNA, it may be a standing wave or a moving wave. Either way it is modulated by the temporal variations of the orientations of the base pairs inside the DNA helix. The size scale of the laser-like beam is a quasi one-dimensional quantum wave structure matched to the scale of the DNA molecule..

Manipulations of the trapped photonic wave structure, themselves, are acts of free will; thinking itself is an act of free will. It is an act that changes the configuration of a piece of the Hyperverse during the whole of its process. The piece of the Hyperverse being changed is the photonic wave inside the DNA. If DNA is the processor, then, because each trapped photonic wave structure resides within one observable universe, the piece of the Hyperverse changed by the thought process itself is highly localized. The changes that occur for one thought process span just one 3D observable universe and are localized inside the DNA molecules of the life form that has that particular thought. In such case, each consciousness on each observable universe gets to think independently of the many other independent conscious processes occurring inside the same corporeal hyper-structure that spans many observable universes. In contrast, if the processor were of a type that manipulates electrons, or chemicals, then because the vortices of those objects span many observable universes, the act of thinking would extend into many nearby universes and couple the thought processes in all of those universes. In such case, there would no individual free will in each universe, despite the existence of quantum mechanics. Thus, the degree to which free will is available depends on DNA providing the central processing functions for organic systems and via its manipulation of trapped photonic wave functions.

The requirements on the processor are distinct from those of any memory circuits that the processor might require. Memory can be a static device or even a chemical device. Memory can be stored in many ways without violating the fact that our consciousness observes the Hyperverse one 3D slice at a time. For example, short term memory could be in the DNA for rapid access, while long memory is stored via chemical methods.

The Nature of Action: A key ingredient in the concept of free will is the ability to decide to act. When that moment arrives, there comes a need to make a change to the universe that is orders of magnitude larger than that required to think. The action requires more energy and the ability to generate much more force than is required to think. It requires amplification of a quantum result to manifest a macro scale change to the configuration of the universe. To enable action, the DNA includes a direct interface between consciousness and the physical body. The DNA includes, directly within it, the genetic sequences required to activate proteins which then direct the corporeal self.

A mechanism in the decision to act amplifies the quantum probabilities. As discussed earlier in the chapter, each corporeal human spans many 3D observable universes. Meanwhile, a decision to act is a separate result of thought in one observable universe. In some observable universes the decision to act may occur, while in others nearby it may not. If the thought process always results in the same decision in all observable universes, then there is no real free will. If the thought process results in different results in different observable universes, then their conflicting decisions result in conflicting directions to the corporeal self. The outcome of that conflict may be, or appear to be, statistical; that is, this conflict may amplify the statistical aspects that we observe in quantum mechanics to the scale of macroscopic existence. That does not mean that the behavior is fundamentally random, but rather that it includes a random component. It could be more of a fractal nature, depending on differences so small and ever changing that we never know the initial conditions that determine the outcome.

As odd as a conflict between the corporeal body and an electromagnetic consciousness may initially seem, it is in fact an observed effect in everyday life. A classic example is the "deer in the headlights" type of behavior, which is a well-known example of this rather common behavior. People frequently act in a manner at odds with their intent. There is no proof that the effect is indeed related to the above mechanism. Yet the data is consistent with the model. That is, the data does not disprove the model, and, for now, that is enough.

Motivations – Least Action and Conservation: Just as the discussion of free will is within scope of the hypervortex model, so too is a discussion of the possible role of the hyperfluid properties of Least Action and Conservation in affecting the goals of the exercise of that free will. Is it, though, metaphysics or simple personification to suggest that life's goal of survival is a manifestation of the fluid's conservation properties? Similarly, does Least Action manifest as a metric that we use to select the path towards our goals? Such questions are not new, nor should the credit go to the hypervortex model for raising the possibilities. An initial purpose of the development of the principle of Least Action, at the time of its development, was to answer such metaphysical questions. It was only later that these questions were decoupled from the development of rigorous mathematics that is now used to exploit the identified principles.

If DNA is the processor operating on trapped quantum wave structures, then a detailed computation of the hyperfluid's performance in that nano-system might begin to reveal answers to the above metaphysical questions. Such a processor would be manipulating the detailed structure of a combined vortex-wave structure. Such structure is largely defined by the principles of Conservation and Least Action in balance with the forces of gravity and electromagnetism at the core of that structure. Conservation is important to the stability of such structures in general. Thus, it is not too far a stretch to suggest that free will combined with conservation manifests as survival.

The possible results of combining free will and Least Action are more varied. The combination might manifest as a tendency to minimize thought thus accepting the first action that might achieve survival. Alternatively the combination might manifest as a tendency towards deep thought to find the path to survival that requires the least action. The manifestation might vary across species. If detailed analysis of photonic wave structures in DNA reveals less than complete answers that raises the possibility that other, undiscovered behaviors of the hyperfluid explain these core motivations of life. Whatever initial relations between life's motivations and the fluid's properties might be revealed, it is also important to scientifically address the possibility that the principle of Least Action is itself a personification of some other principle. The principle of Least Action has led to important and powerful mathematics. Yet, that same mathematics may really result from some other principle.

If the hyperfluid properties of Conservation and Least Action are, even in part, motivations for life, then other questions arise. Is the hyperfluid that omnipresent life force that some spiritual cultures say exists? This is not a quest by me to create or spread any such belief, rather a nod to the possibility that believers in such concepts may wish to learn where in the hypervortex model to seek support for such beliefs. Personally, I am equally likely to support the possibility of Least Action as a new excuse for inaction, as in "The fluid made me not do it!" After all, sometimes the least action is no action.

I do hope that at least some readers share my interest in learning the answers to the questions raised here. I would caution though that it might be folly to expect that the answers will be fully developed from the hypervortex model within our lifetimes. Still sometimes it is sufficient to see a possible path that may, in the future, achieve scientific answers to such questions. The hypervortex model provides such a path, which prior models have not, and that makes the hypervortex model superior as regards this topic.

Actions Impact the Hyperverse: When the corporeal body acts, it still has the challenge of changing the Hyperverse. The action taken by the corporeal body is upon particles, but those particles are each a small part of a much larger vortex. Vortices may extend for light-years in the fourth spatial dimension. It is quantum mechanics that enables the action to have an impact. Quantum mechanics provides the surplus length that vortices each have. That surplus length enables local action to have local impact. The level of resistance provided by the vortex rises only to the level of inertia and not more. Subsequent to the initial response, propagation of the impact occurs along the vortex.

The impact of surplus vortex length on the path taken by a vortex was discussed in earlier sections. In Chapter 8, the example of a surplus vortex length allowing the vortex to form a helical path was used. The helical path provides an especially flexible vortex. Helical structures are used frequently in technology in part for this reason. The coiled phone cord of the traditional wire-line phone is a well-known example. In that case the coil provides both flexibility and stretch. Similar coiling is used to create flexible piping. Twisted pair electrical cables also gain flexibility from the coiling, although in that case the design is motivated by other factors. Without the helical path of the vortex, it is not clear that free will and determinism would exist in the balance that we experience.

The ability for free will to change the Hyperverse raises a question that is often the subject of science fiction discussion. The question is, "If a person makes a change in one 3D observable universe, does it propagate instantly to all the other 3D observable universes, past and future, thus changing someone else's "present"? The question is important because if propagation to the past and future is instantaneous, that violates causality. That is, within a 3D observable universe, we would observe effects before the cause occurred. The question becomes further intriguing because, the hyperfluid shows that certain changes to a vortex, i.e. for those involved in "Spooky Action" propagation along the vortex occurs much faster than the speed of light. The hyperfluid, fortunately, provides an answer that does not violate causality. In particular, the hypervortex model, says that, due to the surplus length of vortices and due to the nature of the vast majority of changes made by free, changes will propagate with the 3D observable universe in which they were made. Because vortices are very soft with regard to velocity changes, those changes propagate slowly. In fact, they propagate with the expansion flow; they do not race into the past or the future. They do not cause spontaneous changes to other observable universes.

Definition of Life: The hyperfluid adds technical detail beyond expectation to the philosophic "I think therefore I am" definition of life. That additional technical detail enables expansion of the philosophic definition of life to provide it in a more scientifically useful form. According to the hypervortex model:

Life is an electromagnetic phenomenon that can observe the Hyperverse one 3D slice at time via fundamentally electromagnetic sensing, manipulate the hyperfluid at a quantum level in the performance of analyses of the impact of alternative actions on the Hyperverse, and, wrapped in a corporeal casing, act to change temporal evolution of the Hyperverse to its interests.

This scientific definition of life embodies the philosophic definition of life, whereas prior scientific definitions generally do not. Traditional definitions have been about the ability to grow and reproduce, which, in comparison, seem inadequate to the task of distinguishing life from other phenomena.

If DNA is the organic processor, i.e. the basis of thought, then all DNA based life forms have a level of consciousness. In such case, consciousness does not require neurons. Neurons allow many cells to coordinate and thus enlarge the collective processing capabilities of the many separate DNA molecules. However, with DNA as the processor, cells without neurons still observe, think, and act. That includes single cell organisms. It also includes every cell in a multi-cell organism, not only those cells located in the brain.

God: Einstein's famous commentary on quantum mechanics was, "God does not play dice with the world". The hyperfluid provides insight into Einstein's issue with Quantum Mechanics and tells us much about how right Einstein was. Einstein's point was that while Quantum Mechanical phenomena are well described mathematically by Schrödinger's Equation, it is wrong to interpret statistical outputs of the equations as a fundamental truth. The hypervortex model mostly agrees with Einstein, though there is room for a residual fundamentally probabilistic aspect in the model. Chapter 8 showed that because vortices have surplus length, they have freedom to deviate from their classically computed path. We saw further, that equations like those of Quantum Mechanics can be generated to describe the possible paths allowed by the surplus length without need to invoke any fundamentally probabilistic aspect in the laws of physics. Thus, in principle the location and speed of every point of a vortex can be well defined. In that sense Einstein was right.

Chapter 8 also showed that while the vortex motion is well defined and non-probabilistic in principle, we lack means to know a vortex's initial position and velocity to any accuracy better than the level of Heisenberg's Uncertainty. Therefore we lack means to compute the precise temporal evolution of vortices. We can only observe one point on any vortex at one time. We also know, by analogy to man's extensive attempts to simulate tornadoes and hurricanes, that the computation of vortex motion is incredibly computationally intense and has accuracy limits. Also, fractal theory indicates limits on the accuracy of such computations. Further, to date no exact closed form mathematical description of the path of a tornado or hurricane vortex over time is known. Moreover, the vortex path will be affected by interactions with all types of waves and other particles nearing any particular vortex of interest. Thus, the statistical mathematics approach of Schrödinger is practical and the interpretation of the results as statistical is correct. It is only the interpretation of quantum mechanics as necessarily statistical that is incorrect.

Importantly, the vortex can be in a mix of states – a mix of modes – despite the incorrectness of the concept of Quantum Mechanics as fundamentally statistical. A plucked guitar string vibrates in a mix of a fundamental mode and multiple harmonic modes, yet the guitar string's motion is strictly classical. In principle, the location and velocity of every point on the guitar string can be computed over time in its mix of modes. Even the dependence of the mix of modes on the plucking of the string can be computed. Similarly, vortex motion can comprise a mix of modes. A vortex can be in a mix of states.

A vortex can be put into a selected state. A vibrating guitar string can be affected such that all modes, but one, are damped away. Further, such damping quickly propagates along the string until the whole of the string is in the one selected mode. Quantum mechanics is useful for computing the mix of states that a set of forces will create in a vortex. Further, as with the guitar string, if one of the modes is selected on the vortex and the others damped away, we can expect that damping to spread along the vortex. It seems that quantum entanglement experiments exemplify this.

Whether a residual fundamentally statistical aspect of quantum mechanics exists depends, in part, on whether vortices take the path that is the exact extremum of the motion. If vortices are free to take a path that deviates within some range of the extremum of motion, then there may be a fundamentally probabilistic aspect of Quantum Mechanics. Similarly, whether a fundamentally statistical aspect of quantum mechanics exists depends, in part, whether the fluid is exactly conserved. If, at times, fluid can be destroyed or created or change phase to help satisfy conservation within particles, then such mechanism may also provide a fundamentally probabilistic aspect of Quantum Mechanics. Such possibilities pose a question regarding what physics determines the phase and eccentricity of a vortex path.

Finally, at the intersection of Einstein's aforementioned comment regarding god and the Flatland allegory are new and interesting questions. The questions all start with the realization that the hyperfluid shows that not only is man not at the center of the universe, man cannot, even with the most powerful technology, observe the whole of existence. Are there observers who can simultaneously observe and even affect a substantial portion of the length of a vortex or a substantial number of observable universes? If so, are there methods for us to observe those observers? If so, what would they look like to us? Can we and they coexist in the same hyperverse or even in the same part of the same hyperverse, and is there some technology we can create to similarly observe and affect multiple observable universes? Thus, finally, regarding god:

"How many observable universes must an observer be able to simultaneously observe and affect to be considered omniscient and omnipotent?"

10. Conclusions and Future Directions

The hypervortex model has been shown here to achieve a compact representation of the universe for current known physics – a representation that fits on one page as shown in the recap that follows this chapter. In contrast, such a summary in standard models would require hundreds of pages if it could even be done, and even then there would be many physical phenomena that it could not describe.

The relationship between the hyperfluid properties and the observable properties of the observable universe can be summarized as follows:

1. Particles are the observable portion of hypervortices in the hyperfluid.
2. Gravity is caused by the variable density of the hyperfluid.
3. Electromagnetism is caused by variations in the fluid's velocity as are inherent to fluids.
4. Strong forces occur when the vortex shape near its axis has anomalies compared to the basic vortex shape that creates the forces that we call gravity and electromagnetism.
5. The weak force occurs when two vortices become so close to one another as to disturb the stability of one or both vortices.
6. Quantum mechanics results from (a) the surplus length of the vortices of particles relative to the shortest path defined by Least Action, and (b) the preference for vortices to be of long length.
7. Pairs of entangled particles are our observation of the same hypervortex as it crosses twice through our 3D observable universe, and the "Spooky Action" that occurs between those entangled particles is a manifestation of a high speed wave traveling along that hypervortex, i.e., an "entanglement wave".
8. Galactic structure occurs despite insufficient gravitational forces within them because each galaxy is itself a hypervortex in the hyperfluid, or rather the portion of a hypervortex that intersects our observable 3D universe.
9. Measurement of the expansion rate of the Hyperverse begins the effort to understand the boundaries of the Hyperverse and the external forces that drive it.

The scope of unification provided by this model spans and exceeds the goals of most or all other unified theories. Also, the hypervortex model provides underlying physical causes for the various equations.

The hypervortex model provides a translation of existing physics into simpler mathematical form. Despite various claims that have been made over time by others that curved space-time coordinates are fundamental to physics, the work has shown that simple Cartesian hyperspace coordinates are valid. Lagrange long ago proved with rigorous mathematics that any coordinate system sufficient to write down the Action of a system is valid. The work here has shown that simple flat coordinates meet the criteria and therefore provide a valid simple alternative to the complexity of curved space-time.

In addition to unifying the basic laws of physics and simplifying the mathematics, the hypervortex model has been shown to encompass much more and without the need for patches and *ad hoc* additions that have been tacked on to other models. The hypervortex model explains the vortex shape of galaxies and the super massive black holes at their centers without need for patching in dark matter. The hypervortex model provides new understanding of the mechanisms that drive the universe and thus provides new meaning and new research directions for understanding and exploring the expansion rate of the universe without need to tack on dark energy. The hypervortex model has scope sufficient to discuss the nature of life, consciousness, and free will at a level of precision that transcends prior physics and metaphysics.

The hyperfluid goes beyond explaining the observed universe; it also provides new insights that allow us to do things previously unknown or thought impossible. The hypervortex model shows great potential to enable new technologies that have previously been, at best, the realm of science fiction. Also, somewhat surprisingly perhaps, many of these new possibilities can occur right here on Earth. We can reap technological benefits from the hypervortex model without needing to visit a black hole or create a wormhole to other stars.

The remainder of this chapter presents some thoughts on methods to complete the model, test it, and to exploit it for new technology. Remember, the model is a framework and the book has emphasized the model's scope and correctness as focused on some of the biggest challenges to other models. Work still remains to fill in many details. Also, work to exploit the model for new technology has just begun; a sampling of science and technology directions enabled or invigorated by the hypervortex model is included in the following sections.

Completing the Model: This work has focused on the unification of aspects of physics that other models have been unable to achieve. In that regard, the work has been successful. Still, much work remains to validate the model against various data and perhaps to adapt certain properties of the hyperfluid accordingly. So far, unification has only required that the hyperfluid have properties of density and velocity. In general, fluids may have additional properties such as surface tension, viscosity, impurities, temperature, dimensionality etc. The work has discussed all of these. Some properties have been omitted from the hyperfluid because data suggests the absence of that property. Other properties are omitted because so far no need for them has been identified.

Now that the hyperfluid has achieved basic unification, a review of each additional possible property of the hyperfluid, in detail, is worthwhile. A key approach to such effort is to match specific hypervortex types to specific known particles. Chapter 8 provides the basic mathematical approach. It specifically noted that if a comparison of such solutions to the observed properties of elementary particles exposes discrepancies, such outcome does not indicate a failure of the model but rather a need to add one or more mathematical terms to the master equation.

An additional aspect of the model to be completed is a formal proof that Quantum Mechanics derives from the unified Lagrangian. Schrödinger did two things either by design or by wonderful accident: (1) He specified the amount of surplus length available for particle paths in Hamilton's work, and (2) He simplified the mathematical challenge by computing statistically averaged deviations from the classical path, rather than trying to compute the exact path and structure of each vortex over time. Formal proof of the relationship between his work and the mathematics in Chapter 8 is work in progress.

Mapping Particles to Hyperfluid Structures: There are only a few elementary particles that are truly stable – electrons and protons. Correspondingly, we observe only a few types of vortices in other fluids. It is reasonable that an electron corresponds to a tornado-like vortex, while a proton corresponds to a hurricane-like vortex. Work is required to confirm that these particles match up with the corresponding vortices. If additional fluid properties are needed such that the computed mass or electric charge in the hypervortex model matches measured mass and electric charge that will help complete the model.

Once the stable elementary particles have been matched to vortices in the hyperfluid, then the unstable and metastable elementary particles may be correspondingly matched to unstable and metastable vortices. Leptons may all match up with tornado-like vortices, while baryons may all match up with hurricane like vortices. Identification of quarks, mesons, etc. as structures in the hyperfluid will further complete the model.

One reviewer of the book noted several questions that are particularly pertinent to an investigation of elementary particles as hypervortices. Several of these questions are restated below:

What type of structure in the hyperfluid comprises neutrinos? That is, if photons are waves in the hyperfluid and particles are hypervortices, what are neutrinos?

In particle models, the Higgs boson generates mass and links particles to an omnipresent field. How does that translate in the hypervortex model?

If particles are all hypervortices, what happens to the stability and structure of the various quarks, protons, neutrons, etc. in nuclei where the hypervortices of each are very close together?

These are good follow up questions that the hypervortex model can likely answer. Also, these questions are notable in that they show that the model is of the type that Dirac wanted. Dirac wanted a model that contained a strong physical foundation to replace the mathematically focused theories that had already become prevalent in his time. He was concerned that theories that were predicated simply on the fact that the math produced correct answers were insufficient. He wanted models to have a mathematical representation; however he wanted the mathematical representation to have much stronger physical foundations than is provided by the standard models. The above list of questions from the reviewer shows that the hypervortex model has a strong physical foundation that people can use for deeper understanding of the world.

Testing the Model: To test the hypervortex model, it has been asked whether there is some experiment for which this model gives a different answer than the standard models. Such tests may exist, yet there may be better tests for the hypervortex model. The goal of the hypervortex model is to unify the existing pieces of knowledge and in the process to grow the scope of our knowledge. Thus, it has no goal of diverging from the predictions of the existing piecemeal theories. A way to test the hypervortex model is to use it to predict something that the standard models have no ability to predict. For example, the hypervortex model could be used to compute the shapes of galaxies as vortices. If the model provides vortices that match known galaxies, that is a success that standard models cannot match. An alternate, very different test would be to compute the speed of entanglement waves and compare that to measurement. Standard models have no ability to do that computation. Similar such tests can be created from the various topics presented in the remainder of this chapter.

Exploring Exponential Gravity: Chapter 8 showed that the hypervortex model provides a gravitational force whose strength is slightly different from what standard models assume. This difference is negligibly small most of the time. The difference becomes large at short distances, which resolves several problems regarding our understanding of gravity. "Exponential gravity", so called because of its specific mathematical form, is not actually new. By the 1980's mathematical treatments showed that exponential gravity can provide results that match all experimental data, including tests for General Relativity. The 4D version of exponential gravity could benefit from some additional verification that its results also match known data, which is ongoing.

Now, the hypervortex model provides both strong physical and mathematical foundation for exponential gravity. Gravity attracts objects like the vacuum near a tornado attracts objects. One can say the gravity well directly attracts matter, or one can say that the gravity well bends light which causes objects to follow a path that moves them into the well. Work not presented in the book further suggests that exponential gravity in a globally flat space is equivalent to Newtonian gravity in Einstein's curved space, with one significant difference. That difference is that exponential gravity satisfies the principle of relativity everywhere including at the center of a vortex, while Einstein's representation is fraught with several problems in that regard. Einstein knew about the issues, and they concerned him. Now, the hypervortex model provides a resolution.

The mathematics in Chapter 8 showed which particles cause gravity and why, which are facts that other models cannot derive and hence must assume. The mathematics in Chapter 8 also showed that exponential gravity in the hypervortex model is additive, .i.e., that the strength of gravity grows in proportion to the mass of the particles. Such fact is also something that other models cannot derive and hence must assume. Such power of the hypervortex model is one of the many aspects that make the model attractive as a unifying model.

This new understanding of gravity may provide a technology opportunity. If a gravity well, i.e., a local region of reduced fluid density attracts matter, a local region of increased fluid density will repel matter. This might provide the basis for a "repulsion beam".

Similarly, a directionally created region of low fluid density between two objects could serve as a "tractor beam". Such are the type of opportunities that arise from the improved physical understanding of our universe that the hypervortex model provides.

An intriguing aspect of gravity in the hypervortex model is the potential to understand what happens to particles that fall into a black hole. While other theories lack the unification to allow real analytic work regarding this question, the hypervortex model provides a solid foundation for such discussion. In particular, the hypervortex model says that all particles are hyperfluidic structures, as is the gravity well. Each particle that has mass delivers a little more gravity well to the "black hole" as it falls into the hole. It seems plausible that the particles will completely "dissolve" into the larger gravity well losing any individual identity, like drops of water entering an ocean. Detailed analysis using the hypervortex model can determine whether this plausible result is correct.

Another intriguing aspect of exponential gravity regards its 4D effects. In particular, in concurrence with assumptions shown in some Star Trek movies, the hypervortex model suggests that sliding our consciousness from one observable universe to another might be facilitated by travel deep into gravity wells. According to the hypervortex model, gravity wells slow down both light and the expansion current, i.e. the flow of hyperfluid along the extra dimension. Inside the gravity well things move slower in that direction than outside the gravity well – the deeper the well and the closer to its center the larger the effect. Beyond the basic question of whether this travel between observable universes can occur, there are many questions, such has: (1) how to validate that it has occurred, and (2) is it reversible, i.e. can one return to their original observable universe.

Investigating Exponential Electric Potential: Preliminary analysis suggests that the electric potential may be exponential rather than the standard form, for the same reasons and by similar derivation as for exponential gravity, Work to be done includes performing formal derivation and then examining all of the impacts that this change may have for the applications in electricity and electronics.

Finding Transient Magnetic Monopoles: Our understanding of electricity, magnetism, and optics is fairly mature in standard models because these three topics were already unified via a fluid model when fluid models were previously the mainstream models. Still, the broader unification provided by the hypervortex model includes improved understanding and improved opportunities for manipulating an aspect of magnetism that has long been mysterious.

Magnets generally exist as dipoles; they have a "North" and "South" pole, which are named in accord with the Earth's magnetic poles. However, despite various experiments, it has not been possible for scientists to find or create a separate "North" or "South" pole. Either of those if found, would be a monopole. There is a place in Maxwell's equations for electromagnetism to include a term for magnetic monopoles; that term is currently set to zero.

The hyperfluid equations show that monopole-like magnetic phenomena can exist for short times; they occur when and where certain density fluctuations of the hyperfluid occur. The mathematical details are discussed briefly in Chapter 8. Qualitatively, the circumstances described by the equations are similar to those observed during magnetic flux disconnects and reconnects seen at the sun. Whether the similarity holds up under scrutiny remains to be determined, which makes it a fertile topic for study. Meanwhile various other models for the occurrence of those disconnects and reconnects exist and continue to be generated because those existing models are not satisfying, even to scientists specializing in solar physics. A comparison of the hypervortex model's predictions for transient magnetic monopoles to the predictions made by other models may provide a test of the hypervortex model.

Detecting New Astronomic Events: Amateur and professional astronomers search the skies for astronomic events with both awe and concern. Examples of such events include meteors, super solar flares, and planetary magnetic field inversions. Any one of these events could cause a major disruption to modern society, or worse. The hypervortex model provides potential new understanding of solar flares in that it provides a modification to the existing equations to explain and address magnetic flux reconnection. The hypervortex model also provides potential new understanding of Earth's structure that may impact our understanding of planetary magnetic field inversions. In particular, just as the galaxy may be comprised partly of particles and partly of a massive vortex, so too may be the Sun and the Earth. If so, that changes our understanding of both the Sun and the Earth. That understanding may provide new ability to predict solar flare and magnetic inversion events. It may also help us to better understand Solar and the Earth cycles.

The new understanding of the galaxy provided by the hypervortex model provides additional potential events of concern to explore. Currently we think of the galaxy as being mostly empty space. We tend to assume Earth is safe as it travels through that empty space. The hypervortex model tells us that Earth's travels within the galaxy are fundamentally within a huge vortex. That implies huge concerns for all types of phenomena that can occur within such a maelstrom. Even while there may be no large mass in the path of our solar system, there may be large invisible changes in hyperfluid density or changes in its ambient flow. As our solar system travels through a transition zone, a region where the ambient density of flow changes, that may affect solar output, shake the planet, flip its magnet field, or change orbits. The hypervortex model allows calculating the structure of the galaxy and thus can enable computation of likely turbulence zones, means to observe them, and their possible impacts. Supporters of standard models may counter that dark matter in their models is of similar nature; however, their models lack predictive capability in this regard.

The possibility that Earth is comprised in part of an astronomic scale vortex may revolutionize our understanding of geology and geophysics. A first step will be to find a way to determine whether this is true and to what extent. Then, if it is, this may change our understanding of internal heating of the planet, the sources of radioactivity from the core, the cycle of ice ages, and more. Of course there are explanations now for each of these, but there always is a current best explanation for everything. Scientists explained the sun as a coal burning phenomenon until the concept of fusion occurred.

Achieving Superluminal Speed: Exceeding the speed of light is of great theoretical interest. It can also be of great practical interest if it can be used on Earth. Two technologically useful methods for superluminal travel are revealed by the hypervortex model. The hypervortex model also suggests a third possible concept; however it requires expanding the scope of the hypervortex model.

In one super-luminal technology, the hypervortex model can help us to use "Spooky Action at a Distance" for superluminal information transmission. As discussed in prior chapters, the hyperfluid shows that entangled particles are connected by a vortex and they can communicate via information flow as oscillations on that vortex. The hypervortex model might be used to determine the maximum transmission distance along a vortex and the propagation speed as a function of various vortex properties such as the vortex type. It may also help determine the number of bits that can be sent between entangled pairs and even a bit rate. Thus, it can help determine the useful operating range and the best choice of particle types for best operation of such superluminal communications channels. Determination of detailed transmission properties may depend on refinements to the hypervortex model as revealed by a detailed study of vortex structure. If for example, the hyperfluid has surface tension, that surface tension will affect the speed of waves along the vortex.

While information may travel along a vortex at superluminal speeds, physical travel at superluminal speeds would use a different mechanism – flow in high speed channels in the hyperfluid. Standard models talk about wormholes, i.e. bending all of space to shorten the distance between two points – what a mess that would make of Earth if we tried to use it here. The hypervortex model indicates that high speed fluid flows can be used on Earth without bending space. We are still a long way from creating and controlling a high speed hyperfluid flow, but one could imagine a small laboratory experiment in the not too distant future.

A third superluminal mechanism is perhaps best described by analogy to man's history of technology for traveling in water. For thousands of years man travelled across the vast expanses of the ocean by floating and by sail. Only, recently did man invent the propeller and the energy sources to power it. The propeller enables man to travel at speeds much higher than allowed by floating and sailing, and it allows travel in directions not possible for sailors (for example travelling directly into the wind). Currently, man floats and sails in the hyperfluid; the basic Newtonian rules of motion are determined by the flow of the expansion current pushing objects like the wind pushes a sail. The real breakthrough for man traveling the Hyperverse will be the discovery of the hyper-propeller coupled to the power to drive it. The hypervortex model, however, has provided no knowledge, to date, of the nature of a hyper-propeller. For example, it is not known at the current time whether such hyper-propeller could be made of hyperfluid or whether such a device would need to be made out of something else – hyper-solid (see next sub-section), or flotsam that may or may not be floating in the hyperfluid. Studies to address such approach to super-luminal travel could generate new specialties in hyper-physics. Also, success, if it enables travel along the direction of the hyperfluid flow, would provide one technology component for time travel.

Finding Hyper-Solid, Hyper-Gas: Fluids generally can become gasses or solids in appropriate circumstances. Some fluids have various and multiple such forms. However, nothing is known about whether the hyperfluid can be made solid or gaseous. Until now, man has not even known to ask the question. The existing data about the universe does not appear to include circumstances in which the hyperfluid becomes solid or gas. Now, new studies in hyperfluid physics could explore whole new material types based on hyper-solids and hyper-gas. If we can make a hyper-solid, then we can make the hyper-propeller. Currently though it is not clear what an experiment in hyper-solids would look like –under what conditions does the hyperfluid become solid, and how does one create those conditions using only hyperfluid? Hyper-gas may exist at the center of vortices. Where the pressure drops at the center of vortices, if the hyperfluid has surface tension or other such property, then, where space would otherwise have become devoid of fluid, hyper-gas may "evaporate" from the fluid form.

Developing Alchemy and Transmutation: Deciding between these two words to describe this technological future has been difficult at best; both include strong connotations of impossibility that must be put aside to discuss such technology. For a time in history, alchemists attempted to transmute some materials into other materials. Some transmutations were possible but others were not – turning straw into gold, a historic favorite goal for alchemists, was not. We learned from that history that some types of materials, such as gold, are immutable elements. Much later we learned that, via the processes of fusion and fission, certain elements could be changed, but requiring much technology, radiation, and risk.

At the time when alchemists learned that transmutation of elements was not possible, even the making of light, i.e. photons, could only be achieved as a byproduct of a natural and destructive process – fire. Since then, thanks to Maxwell's equations for electromagnetism, we have learned to make radio waves with great deliberateness, control, and efficiency. Radio waves are photons, but not in the frequency range that can be detected by our eyes. Alchemists did not even know such things existed. Then, thanks to Schrödinger's equation we learned to make materials that could directly generate optical photons with great control and increasing efficiency, and even reverse the process to convert light directly into electrical energy.

Einstein, of course, told us that light is a particle. Thus, in creating radio waves and light we are directly creating particles to our specifications, something even beyond the goal of alchemists. We now take such capability for granted, yet we still mostly consider the idea of such highly controlled and refined conversion between energy and matter to be impossible.

The hypervortex model seems to provide, in its equations, the fundamental knowledge for creating matter. The difference between making photons and making matter is that matter has mass while photons do not. Mass requires a gravity well. Maxwell's Equations, we have seen, are, approximately, the fluid equations without gravity wells. That is, they are the fluid equations in the approximation that the fluid density is a constant. Thus, they do not provide the knowledge needed to make particles that have gravity wells, i.e., that have mass. The hypervortex model provides the missing equation, the equation that tells us how gravity wells occur in a fluid. The equations also show that the gravitational and electromagnetic energy couple and thereby create the observed quantum uncertainties. The importance of that for transmutation is that it shows that coupling energy into and out of gravity wells is possible – it happens all the time.

We know, in general how to make vortices in fluids. We make them in water. We see them in air as tornadoes and as hurricanes and we can recreate them in small scale. With the new equation added to Maxwell's Equation, some clever technologist may be able to use energy to make vortices in the hyperfluid that stabilize as elementary particles. Over time, with finer control we may be able to make the process selectable for a chosen elementary particle. We may be able to both create and destroy particles and thus be able to convert one form of matter into another – transmutation.

A derivative technology of transmutation may include the making of stable trans-uranium "heavy" elements, elements whose atoms each contain more protons (and neutrons) than in the naturally available set of elements. Such elements can be useful for many technologies. Historically, efforts to make such elements have involved smashing atoms together with the hope that some fragments stick together. There has been some success with such an approach. The hypervortex model, first, shows that stable heavy elements are possible; galaxies are super-heavy stable vortices. Just as with any fluid, there should be a continuum of stable heavy vortices of various sizes from the smallest upwards. We just have to know what they are and have more controllable techniques for creating them. The hypervortex model's equations should allow calculating the spectrum of such vortices, their properties, their masses, charge, etc. That will provide a much improved theoretical foundation for a renewed effort to create stable heavy elements. Then, if we can achieve transmutation, we should be able to generate a variety of such elements.

Life, Death, Free Will and Technology: In Chapter 9, DNA as the central processing unit of life was introduced, as the only known structure consistent with our observation of the Hyperverse as separate three-dimensional universes. This certainly counters the neural theories of thought, yet it does explain how smaller organisms and plants can have thought and hence free will. Given the embedded orthodoxy of the neural theories of thought within the scientific community it will likely take a great while before more experiments are performed to study DNA as a central processing unit (not as a chemistry processor, but rather as a nano-electromagnetic processor.) When that community is ready, some of the questions include:

1. How good is DNA as a trap for electromagnetic fields?
2. What is the computation capacity per unit length of DNA?
3. Are the supposed "junk DNA sequences" between protein sequences actually the computational machinery for activating the proteins?
4. Exactly how is the DNA computer powered?

One related item of interest that may trigger research sooner is the ongoing concern of the effect of electromagnetic fields on life. To date, research tends to focus on whether such fields cause cancer. If the hypervortex model is right, the concern should be that the fields interfere with basic thought. While the fields are present, they may create noise that interferes with basic functions of thought. As soon as the interfering fields disappear normal function can return. 50 and 60 Hertz fields from all power lines (not just the large three-phase overhead power lines) as well as cell phones can create such interference. Fortunately, water attenuates these fields well so that the fields only penetrate a small way into the human brain. Smaller organisms will be more affected.

Another area of related research that might interest some researchers is the possibility of inter-universe communications. This might be of particular interest to those who study ghosts. Ghosts, if they exist, seem to be related to inter-universe information leakage. Ghosts may occur when neighboring observable universes evolve in divergent ways – what if a person is alive in one block of observable universes and dead in another block of observable universes?

A final sub-topic that might initiate some research into the hyperfluid's potential contributions to our understanding of these topics is its impact on the meaning of death. If DNA is the central computer, then does death occur when it loses power? How long can DNA continue to operate when access to power is lost? If operation ceases, can it be restarted? If it is restarted, is it the same being with the same free will decisions?

In Conclusion: Mainstream physics theory development efforts drifted away from fluid models of the universe about 100 years ago. Without that fluid, mainstream theories have been unable to provide a unified theory of everything. The work presented in this book has shown that a hyperfluid in which particles are the observable part of hypervortices provides an excellent framework for a fully unified theory. Such a model restores and modifies the concept of a fluid-filled universe. A fluid-filled universe had been the mainstream view of leading physicists like Descartes, Newton, Maxwell, Kelvin, and Einstein from the 17th century into the 20th century. A community of physicists working together could complete the hypervortex model. Success would likely usher in a new era of even more futuristic technologies than we have achieved to date.

References

This small set of references listed here provides, mostly, additional mathematical backup supporting Chapter 8. The references address the mathematics and physics components reused in the hypervortex model. Thus, they address specific representations of electromagnetism and quantum mechanics. They also address prior work on the hypervortex model. They do not reference general relativity or theories of elementary particles. Such mathematics is not reused in the hypervortex model.

With the exception of the first publication in the list, which is prior work on the hyperfluid, the publication dates of the references are not recent. This reveals the absence of any need for new abstract mathematics in the work. Also, their heritage provides a perspective with some traceability to one popular fluid models of the universe. For example, they are published in the same timeframe when the Higgs boson was first proposed.

The first reference is the prior work on the hypervortex model. It was focused at the time on showing that special relativity works in a flat Euclidean hyper-space with global time. It also includes a first solution to the hypervortex model. In particular, it addressed the overall expansion of the universe.

The second and fifth references are important to the overall mathematical approach for the hypervortex model. Those references discuss the Lagrange methods and related mathematical methods that are used to derive the equations of motion of the fluid. The equations of motion are used to compute specific physical phenomena from the basic unified formulation of the hyperfluid.

The second and third references are important to understand how the representation of electromagnetism derived here matches known representations. The representation used here is not the more commonly known version in terms of electric and magnetic fields. Rather, the hypervortex model uses the vector and scalar potential in a covariant form (compliant with special relativity). Thus these references are important for those trying to verify the hyperfluid representation, but who may be unfamiliar with the vector and scalar potential formalism.

The fourth and fifth references are important to understand how quantum mechanics relates to the equations derived in Chapter 8. In particular, they provide foundations for proving that Schrödinger's equation is a linearized approximation of the hypervortex model's equations.

1. Gary Warren, "Coexistence of Global and Local Time Provides Two Ages for the Universe", [http://xxx.lanl.gov/abs/astro-ph/9912116], (1999).
2. Frederick W. Byron, Jr and Robert W. Fuller, Mathematics of Classical and Quantum Physics, especially pages 40-41 (1969).
3. John David Jackson, Classical Electrodynamics, Wiley and Sons, especially p377-380, (1962).
4. Albert Messiah, Quantum Mechanics Vol. I, Wiley and Sons, p 67 (1958).
5. Wolfgang Yourgrau and Stanley Mandelstam, Variational Principles in Dynamics and Quantum Theory, W. B. Saunders Company, Philadelphia, especially Chapter 7, (1968).

Recap

The one page recap of the mathematics on the next page confirms the conciseness of the essential mathematics for the hypervortex model. The recap includes the one master equation (i.e. the Lagrangian Density), plus the equations of motion that derive from it. It also identifies the coordinate system being used and includes key definitions. All other physics derives from those equations. Keep in mind that the recap could be even more concise in the sense that the equations of motion are themselves derivable from the master equation and therefore could be omitted from the recap.

The equations in the recap may provide only a framework rather than a complete unified theory in the sense that the master equation may require additional mathematical terms that incorporate the effects of lesser properties of the hyperfluid, whose need has yet to be exposed. Such properties may include for example surface tension, a fluid property seen in water. If new terms are added to the master equation, then the equations of motion must be re-derived from the new master equation.

Numerical values for the constants k_G and k_V are expected to be fully determinable from known constants such as the gravitational constant, g, and the permeability of free space, μ_0, among others. The numerical values the new constants will depend on the units chosen for the various constants (e.g. Gaussian units or MKSA units).

Hypervortex Model Lagrangian Density

$$L = \frac{P_\mu^2}{2M} - \frac{k_V}{2}\left(\frac{\partial P_\mu}{\partial x_\nu} - \frac{\partial P_\nu}{\partial x_\mu}\right)^2 - \frac{k_G}{2M}\left(\frac{\partial M}{\partial x_\mu}\right)^2 + \lambda\left(\frac{\partial M}{\partial t} + \frac{\partial P_\mu}{\partial x_\mu}\right)$$

Equations of Motion

$$\frac{k_G}{M}\frac{\partial^2 M}{\partial x_\mu^2} - \frac{k_G}{M^2}\left(\frac{\partial M}{\partial x_\mu}\right)^2 = \left(\frac{\partial \lambda}{\partial t} + \frac{P_\mu^2}{2M^2}\right) \qquad \textit{(1 equation)}$$

$$2k_V\frac{\partial}{\partial x_\nu}\left(\frac{\partial P_\mu}{\partial x_\nu} - \frac{\partial P_\nu}{\partial x_\mu}\right) = \frac{\partial \lambda}{\partial x_\mu} - \frac{P_\mu}{M} \equiv \frac{4\pi}{V_R}J_1 \textit{ (4 equations)}$$

$$\frac{\partial M}{\partial t} + \frac{\partial P_\mu}{\partial x_\mu} = 0 \qquad \textit{(1 continuity equation)}$$

Coordinate System

$$ds_{Euclidean}^2 = dx_\mu^2, \qquad \mu = 1,2,3,4 \textit{ (or more)}$$

Definitions

$L \equiv$ *Lagrange Density*

$P_\mu \equiv$ *Hyperfluid momentum, a vector field, over x_μ and t*

$M \equiv$ *Hyperfluid density, a scalar field over x_μ and t*

$\lambda \equiv$ *Conservation Field, a Lagrange Multiplier, a scalar field over x_μ and t*

$k_G \equiv$ *Gravitational Energy Constant, proportional to g of standard models.*

$k_V \equiv$ *Vector Potential Electromagnetic Energy Constant,*

$V_\mu \equiv$ *P_μ/M; Hyperfluid velocity, a vector field over x_μ and t*

$V_R = R/t$; *Observable Universe Inertial Expansion Rate, as derived.*

$\rho \equiv \frac{V_R}{4\pi}\left(1 - V_\mu^{-1}\frac{\partial \lambda}{\partial x_\mu}\right)$; *Charge Density, a scalar field over x_μ and t.*

$J_\mu \equiv \rho V_\mu$; *Current Density, a vector field over x_μ and t.*

www.ingramcontent.com/pod-product-compliance
Lightning Source LLC
Chambersburg PA
CBHW021558210326
41599CB00010B/494